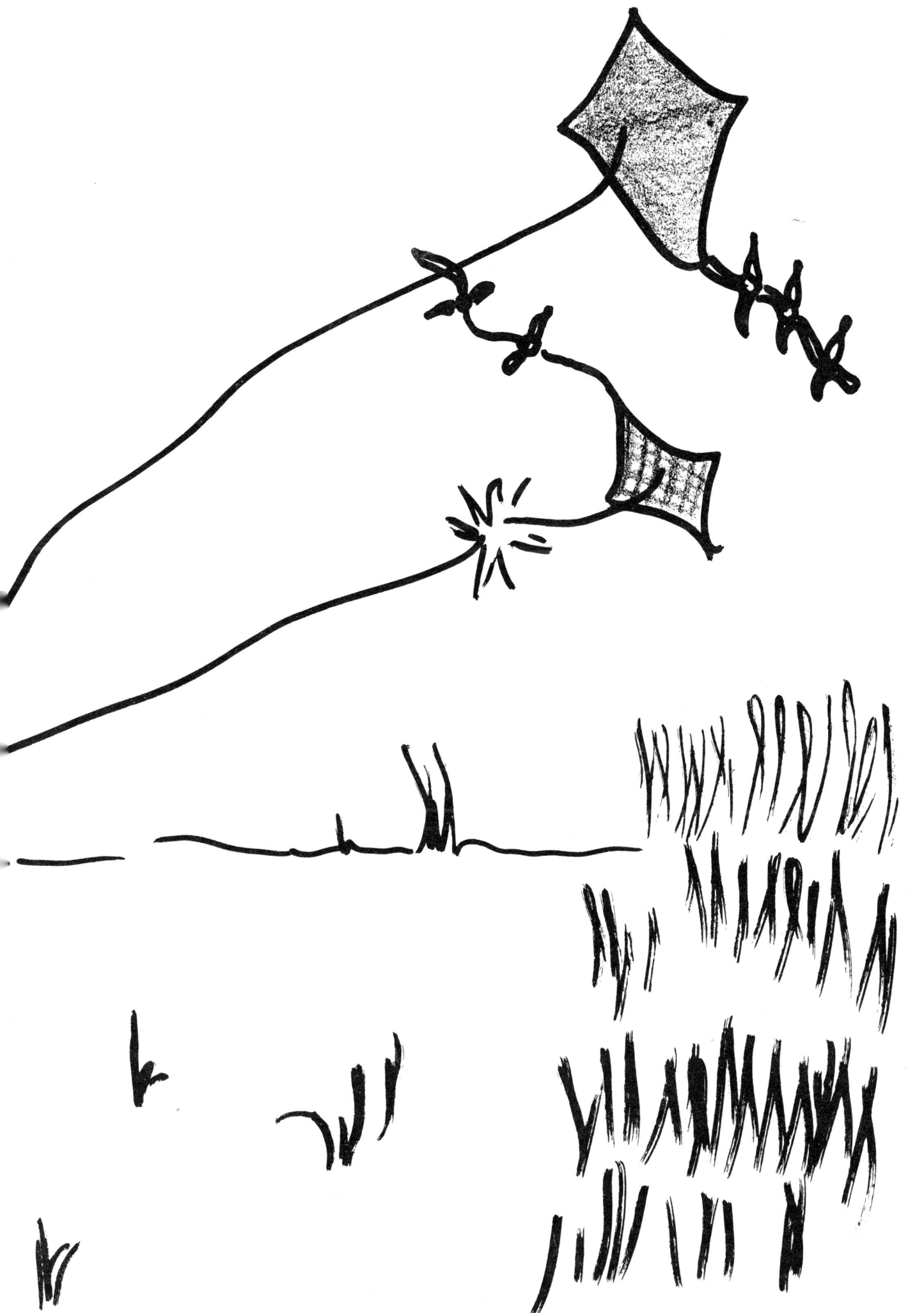

Copyright 1973 by Incentive Publications, Inc. All rights reserved. No part of this publication may be reproduced, stored in a retrieval system, or transmitted, in any form or by any means, electronic, mechanical, photocopying, recording, or otherwise, without prior written permission of Incentive Publications.

4th PRINTING

Printed in Nashville, Tennessee
by Incentive Publications
Box 12522
Nashville, Tennessee 37212

SECOND EDITION

372.35
F776c

# HOW TO USE THIS BOOK

*(Instructions for parents and teachers)*

*Young children are curious about and extremely sensitive to their environment. They instinctively push and pull; take apart and attempt to put together again; smell, taste and feel things around them. "Why", "what", "when", "where" and "how" are words they use naturally and often. It is this interaction with their environment that grown-ups can either nurture and encourage or inhibit and retard.*

*Time to question, time to explore, time to wonder and to ponder is basic to the creative nourishment of the scientific attitude in young children. As they discover each new scientific marvel they should not be pushed to abandon an interest before their natural curiosity has been satisfied. As children are provided with time and help to see relationships and arrive at satisfactory conclusions based on their own observations, explorations and questioning, the foundation for scientific thinking and action in later life is being built.*

*At no time in his life is it more important for the child to understand the world in which he lives and his own role in it. It is at this developmental level that he is building the feelings about himself that will be with him for the rest of his life. The security that is to be gained from understanding of conception, birth and death as natural and normal, and from acceptance of himself as a biological organism with body needs comparable to all like organisms are reassuring to the young child. These understandings can be broadened to include proper health and safety measures as well as to enhance the development of a satisfying and expanding self concept.*

*Exposure to television, radio and adult conversation enable today's children to accumulate an awesome body of scientific knowledge long before they are ready for school. Oftentimes they form false impressions or misunderstood information gained in this manner.*

The purpose of this book is to give children concrete experiences designed to help them sort out and make meaningful use of this scattered knowledge and isolated concepts as well as to develop new insights and understandings.

The activities have been structured to encourage an open and creative approach to problem solving and to result in the development of broad major concepts of a scientific nature. They have been planned to correspond to the specific interests of the creative pre-school child. Their goal is to encourage the child to learn to form generalizations that will be helpful to him as he expands his horizons and moves on to new levels of curiosity and greater interest in and understanding of himself and the world in which he lives. The units may be presented to the child in any order desired so long as the entire unit is treated as a whole and completed in the proper order before another is started. The activities within each unit are developmental in nature and should be completed in the sequence in which they are arranged.

Many opportunities to explore the wonder of nature first hand; to experiment in the kitchen bath room and back yard with "real materials"; to gain sensory images in feeling, seeing, touching, tasting and smelling, and to verbalize in meaningful discussion about these experiences will enforce the use of this book immeasurably. As the child is lead into paths of divergent thinking, is presented with more than one alternative as a possible solution to his problem and as he internalizes the concepts being gained from observation, experimentation, and classification he will be developing the desired scientific attitude - more importantly, he will be a busy, creatively engaged youngster capable of enjoying each day of his life to the fullest.

## ACKNOWLEDGEMENTS

We gratefully accord the special acknowledgement that is due Miss Paula Oldham whose delight in living, unique artistic flair, and enthusiastic committment to the project have complemented the authors' efforts in the truest sense.

Appreciation is also extended to her journalism teacher at Lynchburg Baptist College, Miss Marie Chapman, whose encouragement and support made Paula's participation possible.

Our sincere thanks to Mrs. Carla Nading who assisted in the typing and preparation of the final manuscript.

CONTENTS

I. Living Things . . . . . . . . . . . . . . . . . . 1

II. Earth and Sky . . . . . . . . . . . . . . . . . .67

III. Water and Air . . . . . . . . . . . . . . . . . .91

IV. Machines, Magnets, and Electricity . . . . . . . . . . . . 105

V. The Human Body . . . . . . . . . . . . . . . . . 127

VI. *Test Yourself* . . . . . . . . . . . . . . . . . 154

All living things need air, sunlight, and water to grow.

Some living things are plants.

Some living things are animals.

Make an X on the pictures of living things.

Color the plants.

# What is a plant?

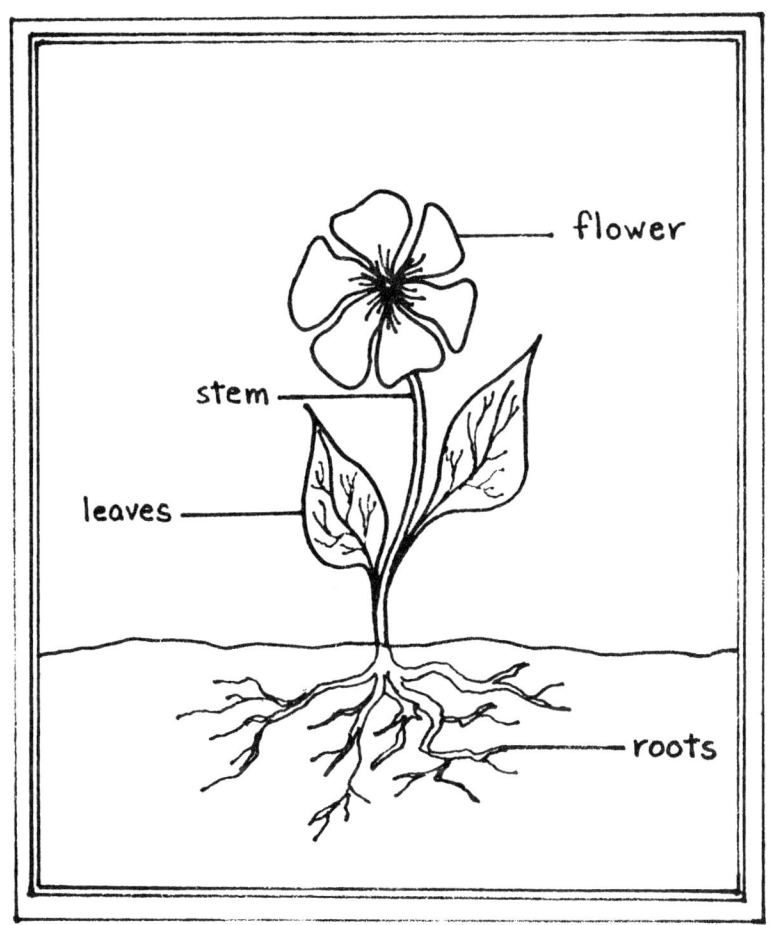

Most plants have four parts.

They are roots, leaves, stem, and flowers or fruits.

Color this plant and say the parts.

Find a real plant and name the parts.

Draw a plant you would like to grow.

Be sure it has roots, stem, leaves and flowers or fruit.

MY SPECIAL PLANT

by _____

Here are some plants with missing parts.

Make the missing parts.

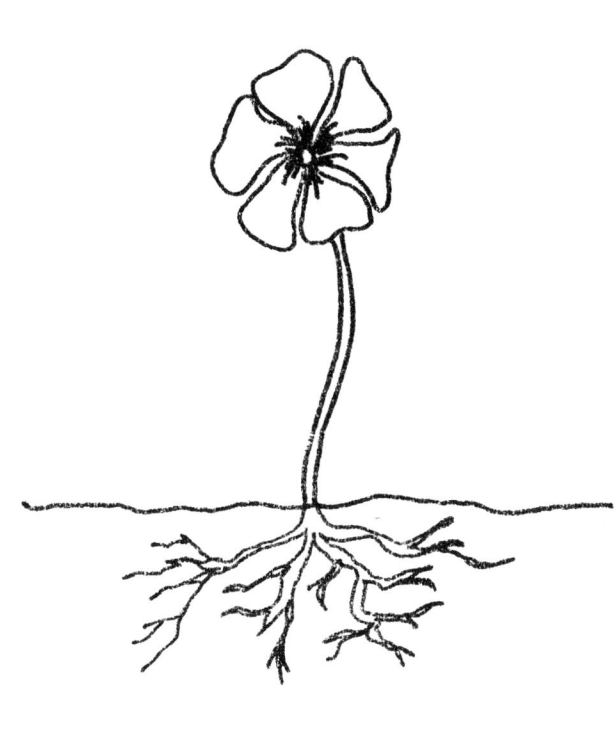

Name the parts.

Color the plants.

# How do plants grow?

The roots of plants grow downward.

The stems grow upward.

Leaves and flowers grow from the stems.

As they grow they turn toward the sunlight.

Make some plants grow in this garden.

Did you remember to turn them toward the sunlight?

# Can you be a plant doctor?

This is a happy, healthy plant.

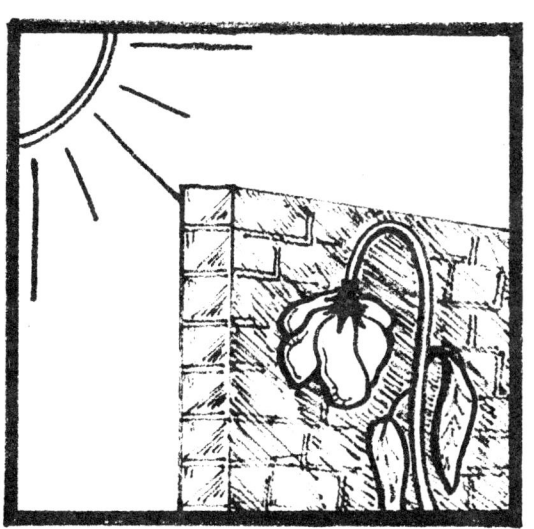
This is a sad, sick plant.

Plants need three things to grow up to be happy and healthy. They need sunlight, water, and air.

Draw a healthy plant here.

# Do you know what happens to plants if they do not get what they need?

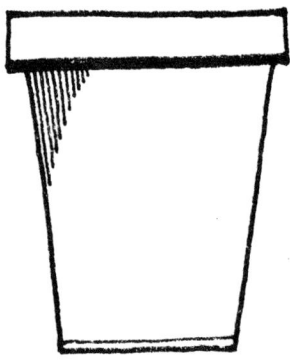 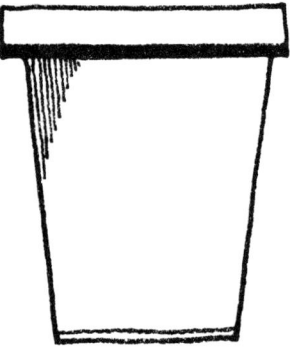

Draw a plant in this pot that has been growing with sunlight, water, and air.

Draw a plant in this pot that has been growing without sunlight, water, and air.

Look around you for plants that are growing with plenty of sunlight.

To find some that do not get plenty of sunlight you might look under big trees.

Mary, Mary quite contrary
How does your garden grow?
With silver bells and cockle shells
And pretty maids all in a row.

# How does your garden grow?

Have you ever seen a garden?

A garden is a place where many plants grow together.

It can be a flower garden or a vegetable garden or a garden with all different kinds of plants.

What kind of a garden do you think this is?

# Can you be a gardener?

The gardener plants and takes care of the garden.

Make your flower garden here.

Put your favorite flowers in it.

Color them pretty colors.

# Have you ever seen a garden you can eat?

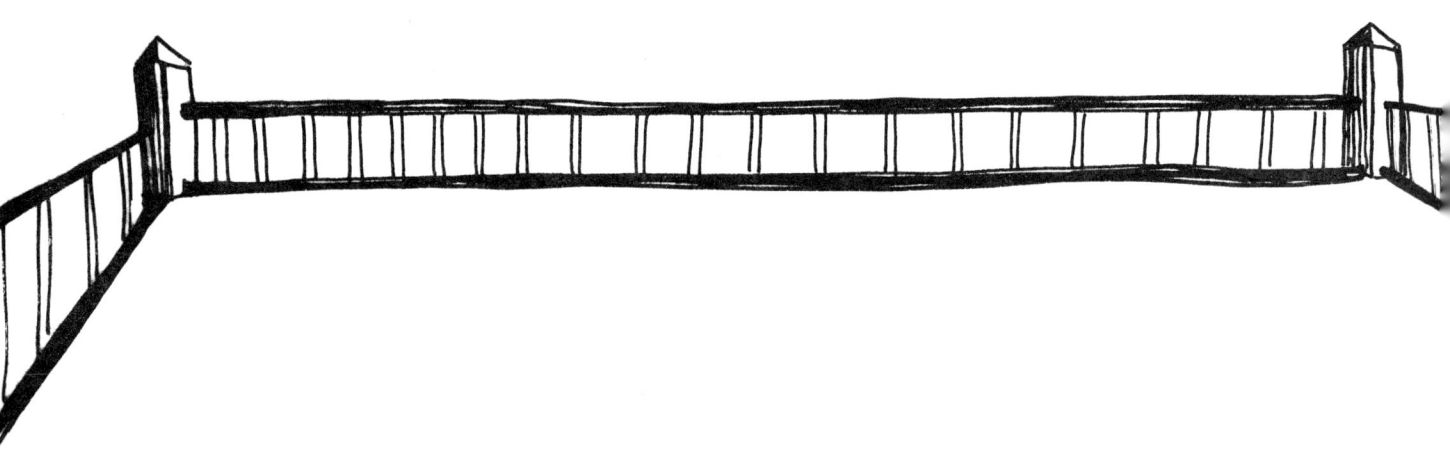

You can eat vegetables from a garden.

Make vegetables you like to eat in this garden.

Say the names of the vegetables.

How are the vegetables different from flowers?

# We eat many different parts of plants.

Make an X on the roots we eat.

Color the leaves we eat green.

Make a circle around the fruits we eat.

Name the stem we eat.

# Where do plants come from?

Some plants can be started from cuttings.

A cutting is a piece of stem taken from an old plant.

It is put in water to grow roots and then planted in soil.

Geraniums can be grown from a cutting.

Can you name another plant that can be grown from a cutting?

 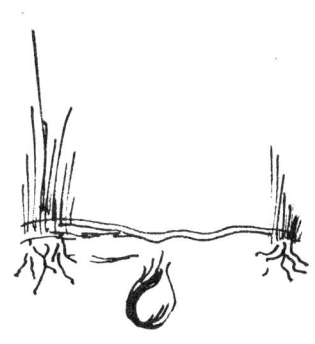 

Some plants can be grown from bulbs.

A bulb looks like a ball of roots covered with tiny dried up leaves.

It is really an underground bud covered with bud scales.

When you plant the bulb in the ground a new plant comes from it.

Tulips grow from bulbs.

Can you draw a picture of another plant that comes from a bulb?

flower

apple

watermelon

acorns

dandelion

Many plants grow from seeds.

It seems strange to think of large trees growing from tiny seeds. They do though.

Can you think of three plants that you know which grow from seeds?

These are pictures of things that grow from seeds.

Can you name the pictures?

FOR YOU TO DO...

Have you ever found the magic star in an apple?

The picture shows you how.

Cut an apple in half as shown.

On each half you will see a star with five points.

Find the seeds in the star.  Count them.

It is fun to pretend the seeds are magic because they will start new apple trees if they are planted.

Scientists "collect" and "classify" things.

This means that they find things and put those that are alike together.

Collect some seeds and classify them
by pasting them together on the next page.

# SEEDS

collected by _____

FOR YOU TO DO . . .

Cut the top off two milk cartons.

Fill the cartons with soil.

Plant a bean seed in each.

Water one carton.

Do not water the other carton.

Watch to see what happens.

Will both plants come up at the same time?

Put the plant that has had water in the sun.

Put the other plant in a dark closet.

What happens to the seed that has the sun and the water?

What happens to the seed that has had no sun and water?

# Would you like to be a scientist?

Scientists keep records of their experiments.

This page is for you to draw a record of how your bean plants grew.

### Seed with sun and water

### Seed without sun and water

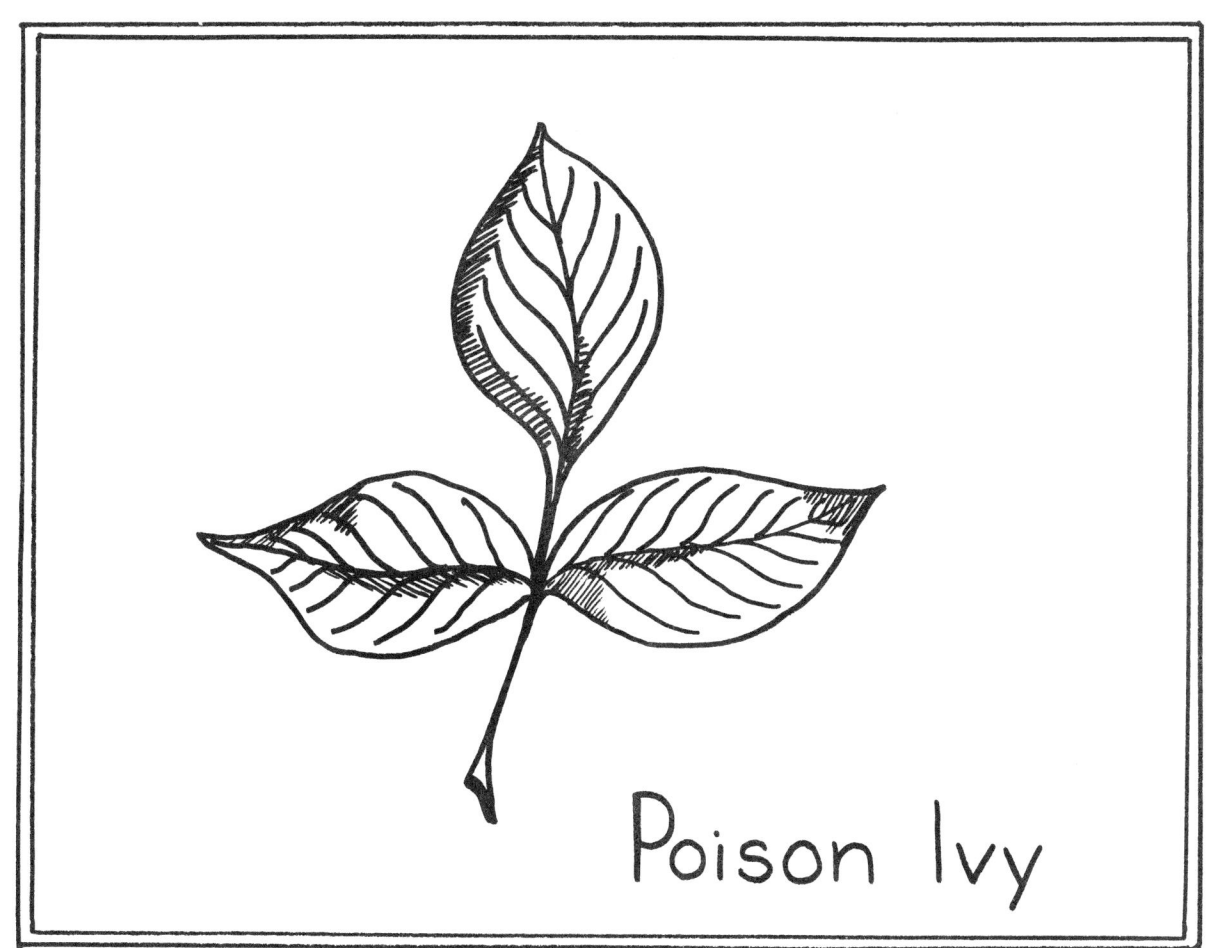

Some plants are harmful to man.

Poison Ivy is a harmful plant.

Do you play with poison ivy?

    Yes _____  No_____

If you marked No, you are smart.

If you know why poison ivy is harmful you are a smart scientist.

Color the leaves of the poison ivy green.

# Some plants have to have special kinds of places to grow.

This is a plant that grows only in the dese[rt.]

It is a cactus plant.

This plant grows only in water.

It is a water lily.

Color the plants.

Different kinds of tools and machines are used to care for plants.

Make an X on the tools and machines that the farmer uses in his fields.

Do you know the names of all these tools and machines?

Make an X on the picture of the tool that the farmer would not use to care for plants.

Make an X on the picture of the tool that the gardener would not use to care for plants.

Make an X on the picture of the machine that would not be used to care for plants.

Make an X on the picture that is not a fruit.

  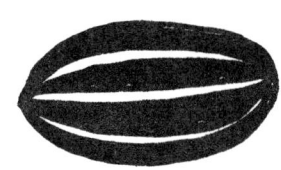 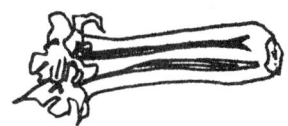

Make an X on the picture that is not a vegetable.

Make an X on the picture that is not a flower.

These pictures tell a story about how an apple tree grows.

Color the pictures.

What colors will you need?

Make a pear tree as it would look with bare branches in winter.

Make a pear tree as it would look with flowers in spring.

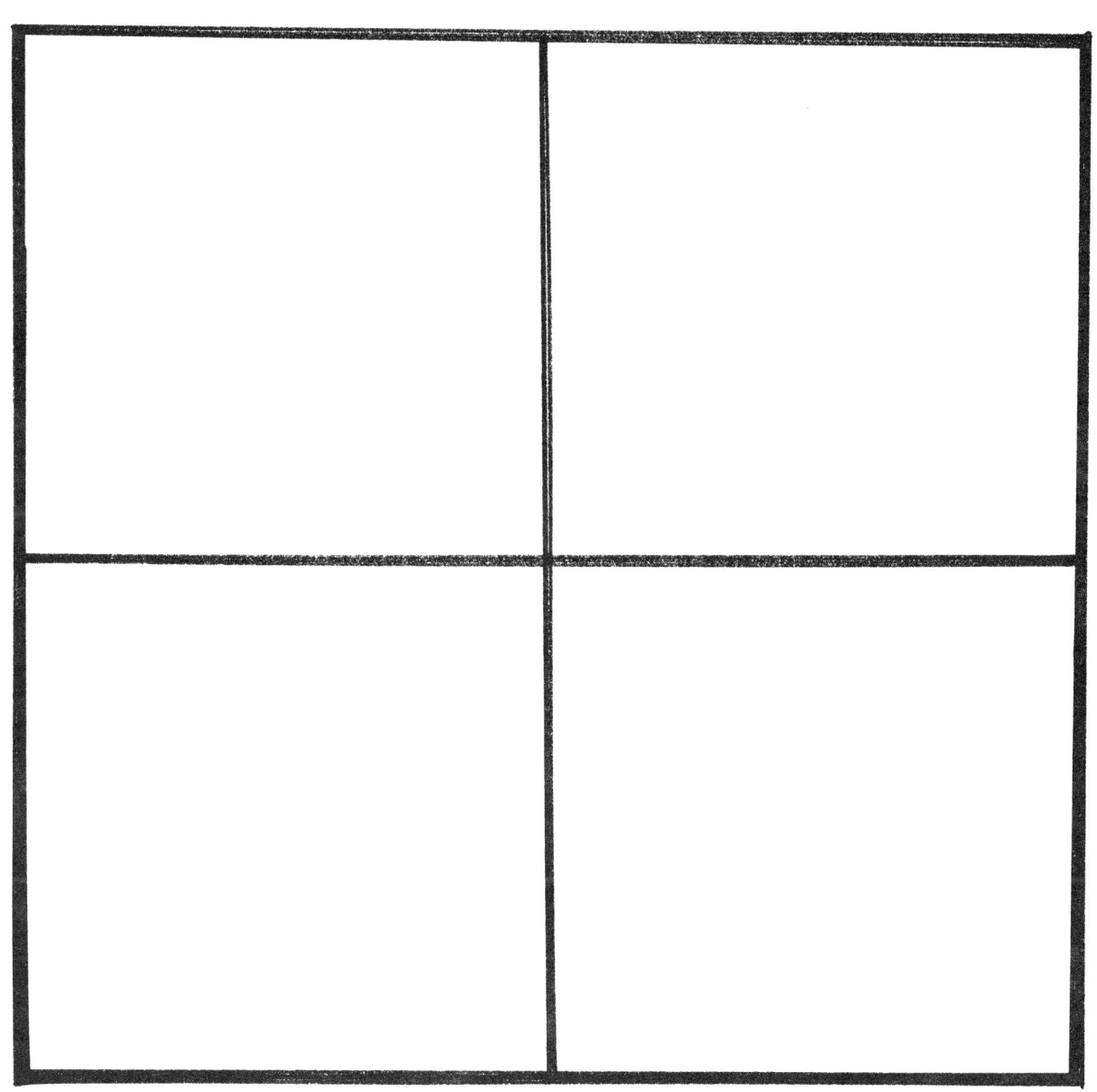

Make a pear tree as it would look with fruit in the summer.

Make a pear tree as it would look when the fruit and leaves are falling off in autumn.

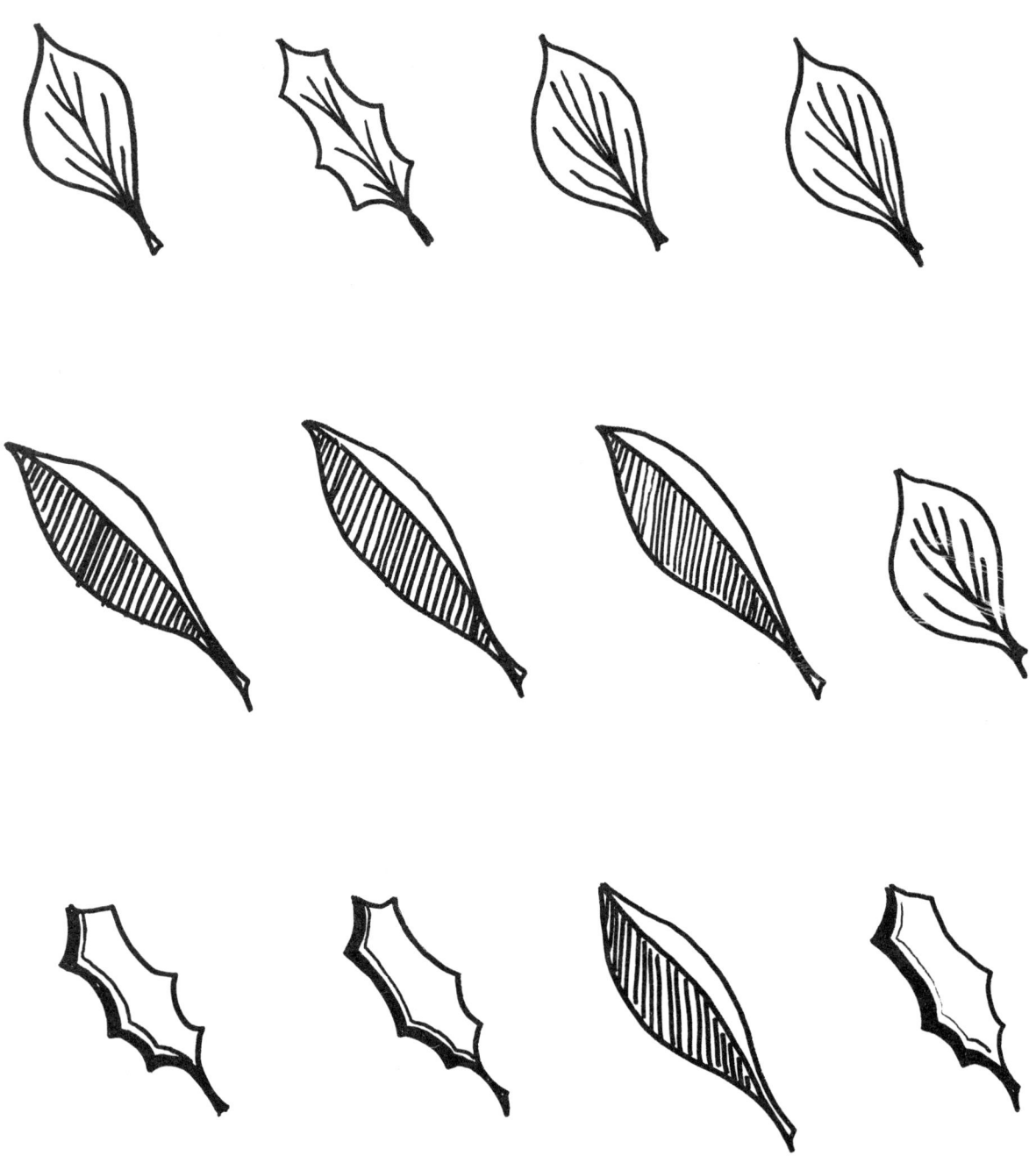

Color the leaves the way they look in the autumn.

Make an X on the leaf in each row that is different from the others.

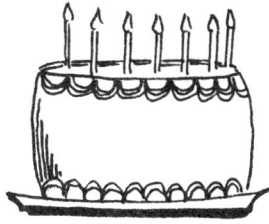  

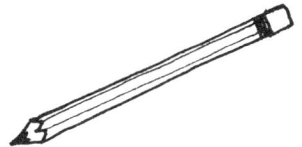  

Make an X on the things that come from plants.

Color the things that do not.

FOR YOU TO DO . . .

    Find a pretty leaf.

    Place it under this page.

    Make sure the underside of the leaf is up.

    Hold the page firmly and color over the leaf.

    Isn't the leaf print pretty?

FOR YOU TO DO . . .

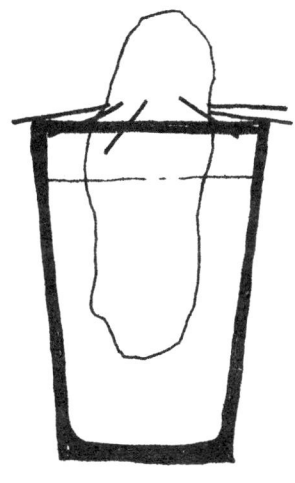

Put one end of a sweet potato into a glass. Stick toothpicks in the potato to hold the top half out of the water.

Put the top of a carrott in a small dish of water. Place some small pebbles in the water.

Watch a pretty plant grow!

Watch another pretty plant grow!

# Where do animals come from?

Some animals are hatched from eggs.

Some of the mothers of babies hatched from eggs get food for their babies. Some do not.

Can you name some animals that are hatched from eggs?

Birds are hatched from eggs.

Mother birds feed their babies when they are young and teach them to fly.

Do you know what mother birds feed their babies?

Color the birds blue.

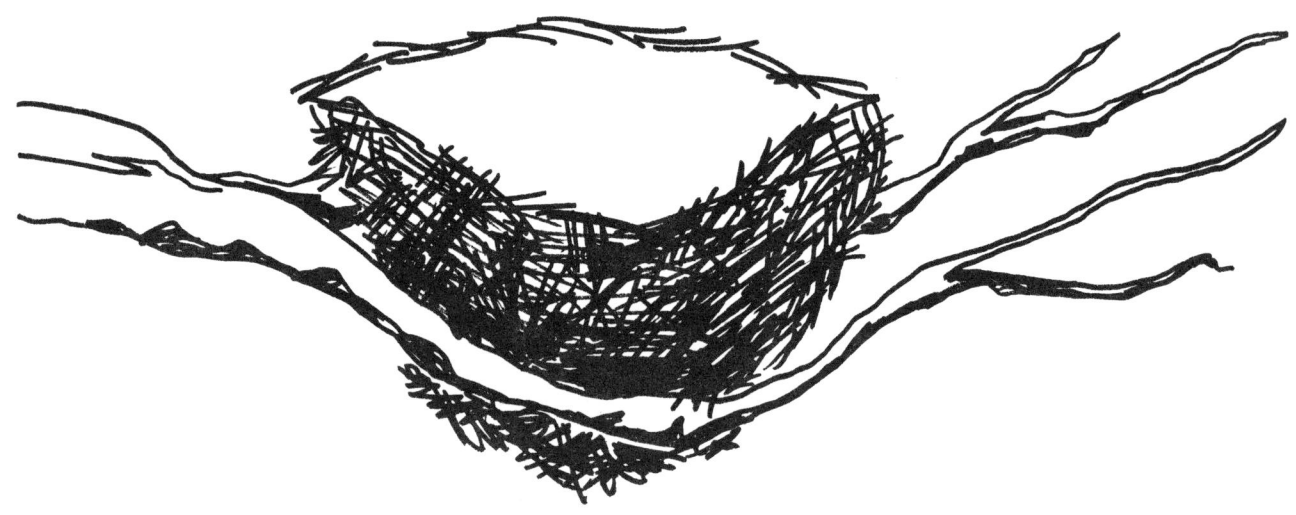

Mother birds build nests for their eggs.

These nests become the baby birds' first home.

Do you know what bird nests are made from?

Draw a mother bird and some baby birds in this nest.

Some baby animals are born alive.

Before birth they are carried inside their mothers in a pouch or bag called the placenta. Scientists use the word mammal for these animals.

All mammals have hair and feed their babies milk.

Write the word *mammal* so that you will remember it.

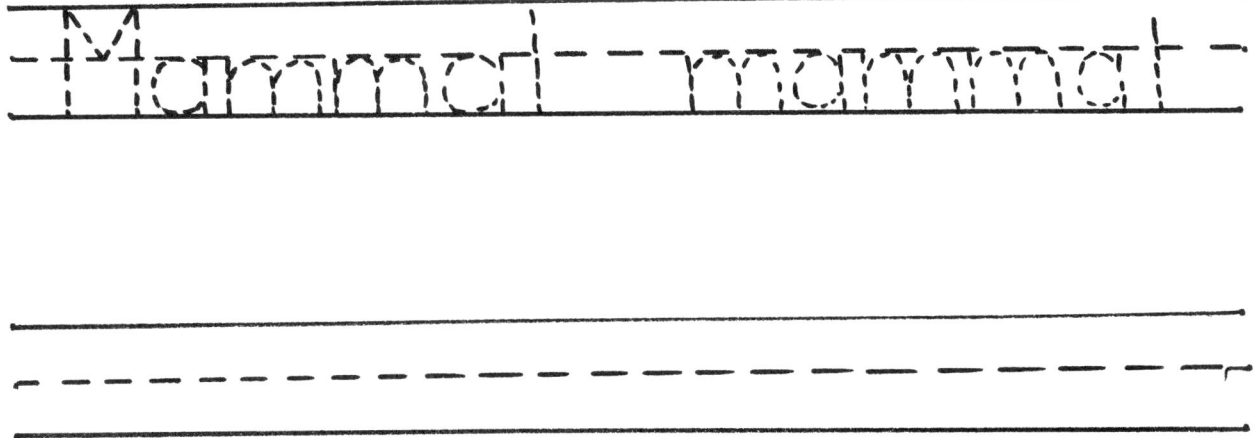

Cats are mammals.

Baby cats are called kittens.

Mother cats give milk for their babies.

Can you think of some other mammals that are born alive and get milk from their mothers?

# Did you know that baby animals are not always called by the same name as their mothers?

Follow the lines to make the animals.

Color them.

What is the name of the mother animal?   Write it.

cow

What is the name of the baby animal?   Write it.

calf

Was the baby hatched from an egg, or born alive?

What will the baby animal eat?

_____ is a mammal.

Did you know that you are a mammal?

Make your picture here and write your name.

# Have you visited a farm?

Here is a barn and a barnyard.

Draw some animals for it.

Do you know what the farmer gets from each of the animals you drew?

Write each of the words.  Say them.

Each one is the name of something we get from one of the farm animals.

Draw a line from the picture to the word it matches.

# How many farm animals do you know

Circle all the farm animals.

Color the farm animals.

Name the other animals.

Pets are animals that live in homes with people.

Make an X on the animals that would make good pets.

Color the others.

Do you know why an elephant would not make a good pet?

Draw a pet you would like to own.

# My Pet Story

by _____

Pets need to be cared for by people.
How would you care for the pet you drew?
Make a story about it.
Ask someone to write it for you.

In each row one of the pets is different from the others.
Make an X on the ones that are alike.
Color the one that is different.

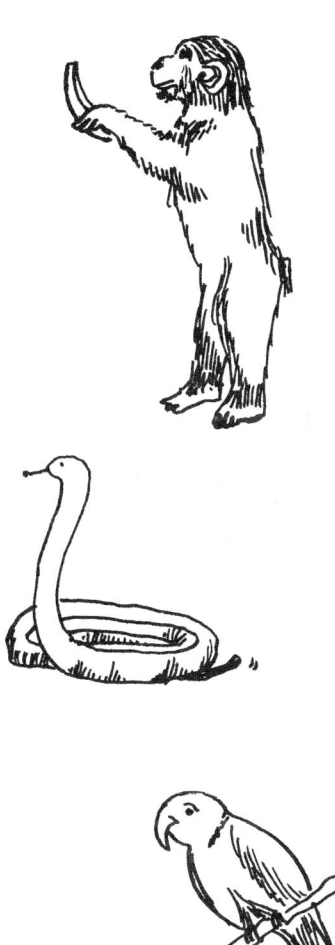

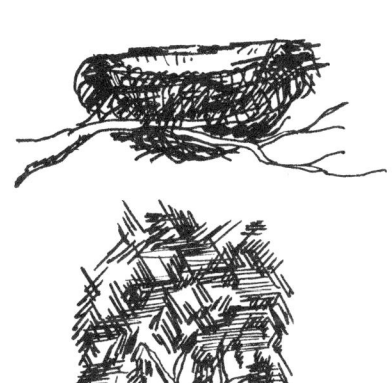

Some animals live in the zoo.
The zoo keeper tries to make the animal's new home as nearly like his old one as possible.
Do you know what kind of homes the animals come from?
Draw a line from the picture of each animal to the picture of its home.
Say the name of each animal and the name of each home.

Insects are animals.

All insects have three main body parts and six legs.

Sometimes we call them bugs.

Which word do you think scientists use?

Write it.

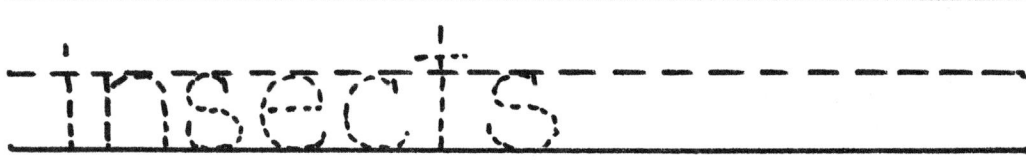

Some insects are helpful to man.

Some insects are harmful to man.

The honey bee is a helpful insect.

Write the name of the food we get from the honey bee.

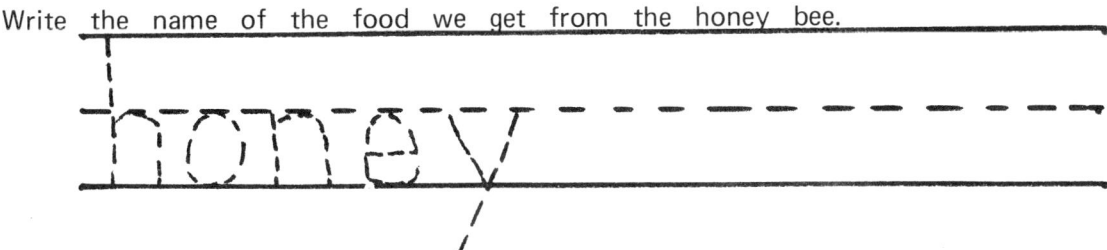

Ask someone to tell you the story of how it is made.

# Why was Little Miss Muffet frightened?

Little Miss Muffet

Sat on a tuffet

Eating her curds and whey.

Along came a spider

And sat down beside her

And frightened Miss Muffett away.

MOTHER GOOSE

A spider looks very much like an insect, but it is not an insect. It is called an arachnid. Arachnids have bodies that can spin tiny threads. Often, they use these threads to weave webs. Some spiders are helpful to man. Others may be harmful.

# A HARMFUL INSECT

by _____

Can you name two insects that are harmful to other animals?

Draw a picture of one of them.

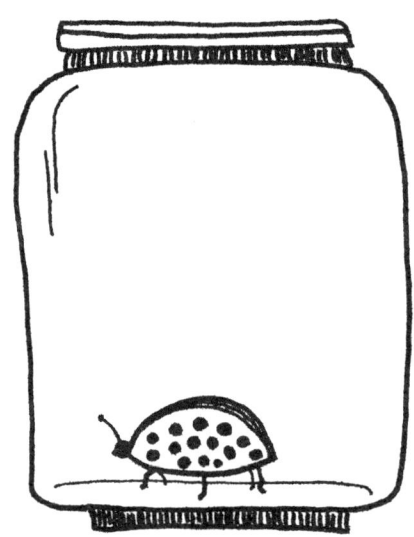

Ask someone to help you catch an insect and put it in a glass jar.

<u>Observe</u> the insect carefully and decide upon three special things you could tell about the insect.

Ask the same person to help you write the story on the next page.

Then let the insect go.

# About My Insect

by _____

Animals have different ways of moving.

Some hop, some crawl, some fly, some swim, and some walk.

How many of these ways have you <u>observed?</u>

Color all the pictures of animals that fly.

# Do you know what animals eat?

Some animals eat plants.

Some animals eat other animals.

Follow the dots to find an animal that eats both plants and other animals.

Some animals use their coverings for protection.

Turtles pull their legs and head into their shell if they are touched.

Snails can pull their whole bodies into their shells.

Have you ever touched a snail?

How many different kinds of animal coverings can you think of?

Draw a line from the picture of each animal to the picture of the thing his covering gives us.

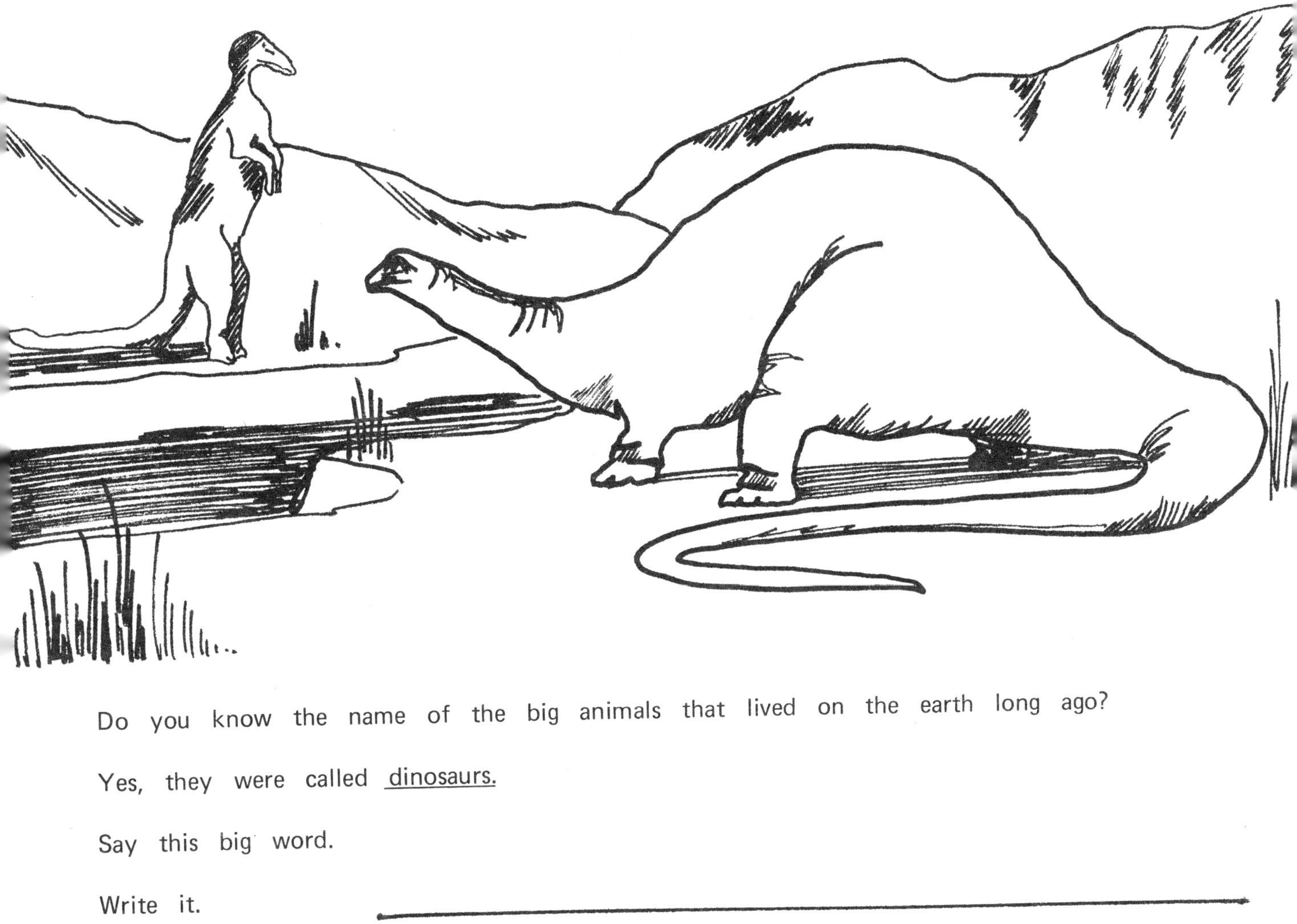

Do you know the name of the big animals that lived on the earth long ago?

Yes, they were called <u>dinosaurs.</u>

Say this big word.

Write it.

dinosaurs

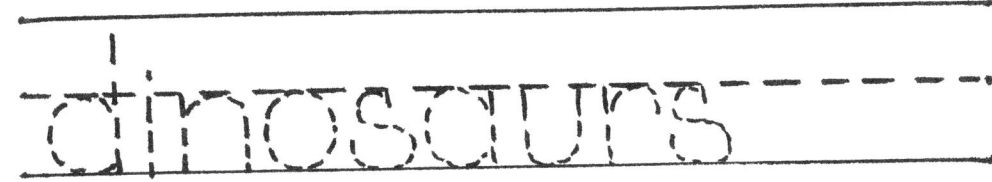

Do you know why the dinosaurs are not living today?

Plants and animals are living things.

They need each other to keep the balance of nature.

Make an X on the picture that is not a plant.

Make an X on the picture that is not an animal.

Make an X on the picture that is not a living thing.

In what ways are these three things alike?

In what ways are they different?

Show the pictures to a friend. See if your friend can help you think of some more ways they are alike or different.

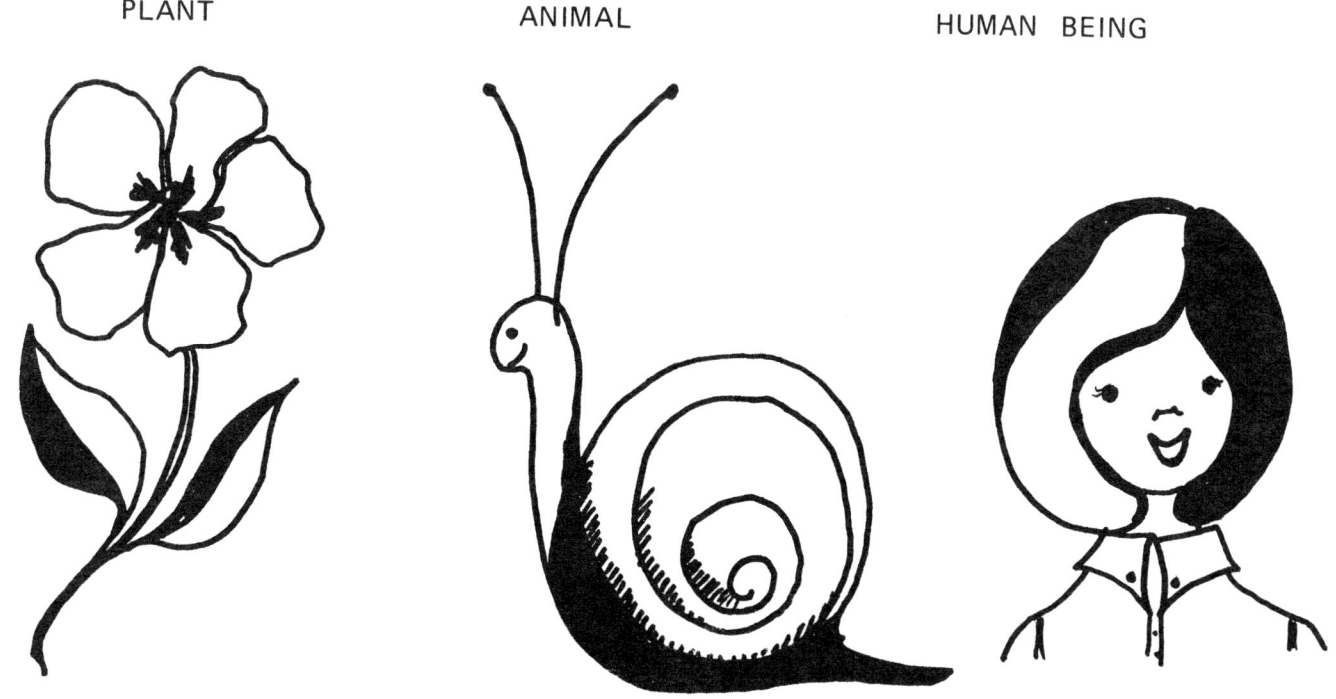

PLANT        ANIMAL        HUMAN BEING

PLANT　　　　　　　ANIMAL　　　　　　　HUMAN BEING

 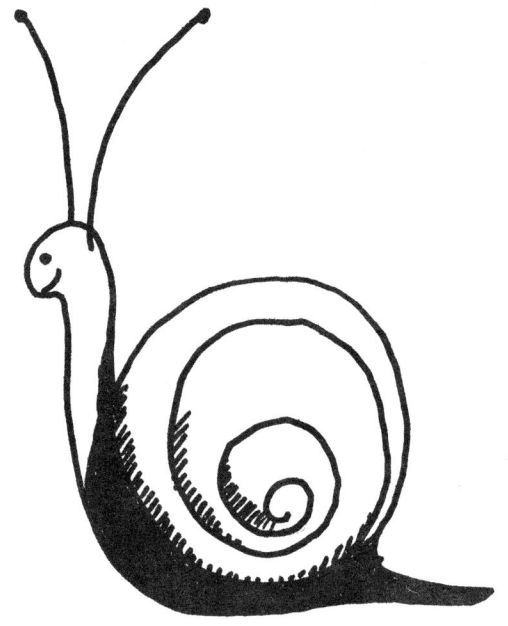 

The pictures are labeled as a scientist would name them.

Can you write the correct name for each picture?

Did your answers match these?

Write them.

The 🌸 is a _____

The 🐌 is an _____

The 👧 is a _____

# What's wrong with these pictures?

Be an explorer! Take a walk in a field or a park, or on a beach. Look for living things. Maybe you will see a butterfly or a spider's web. If you are lucky you may hear a mockingbird sing or smell a honeysuckle in bloom.

# EARTH AND SKY

# Did you know that the earth is round?

It is covered with water, soil, rock, plants and animals.

The shapes show where the land is.

The rest is water.

Color everything except the shapes blue.

Did you notice that you colored more than one-half of the earth blue?

That is because most of the earth is covered with water.

Can you think of some plants and animals that live in water and could not live on land?

Make a picture of two things that live in water only.

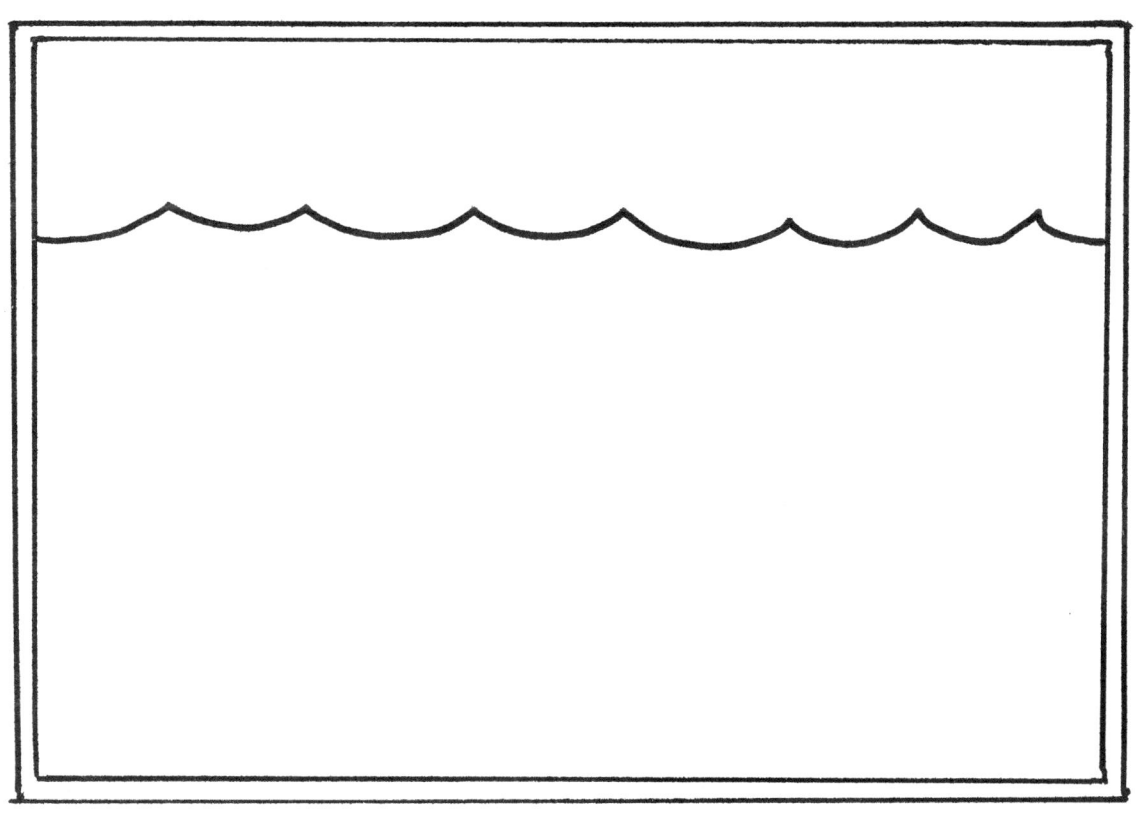

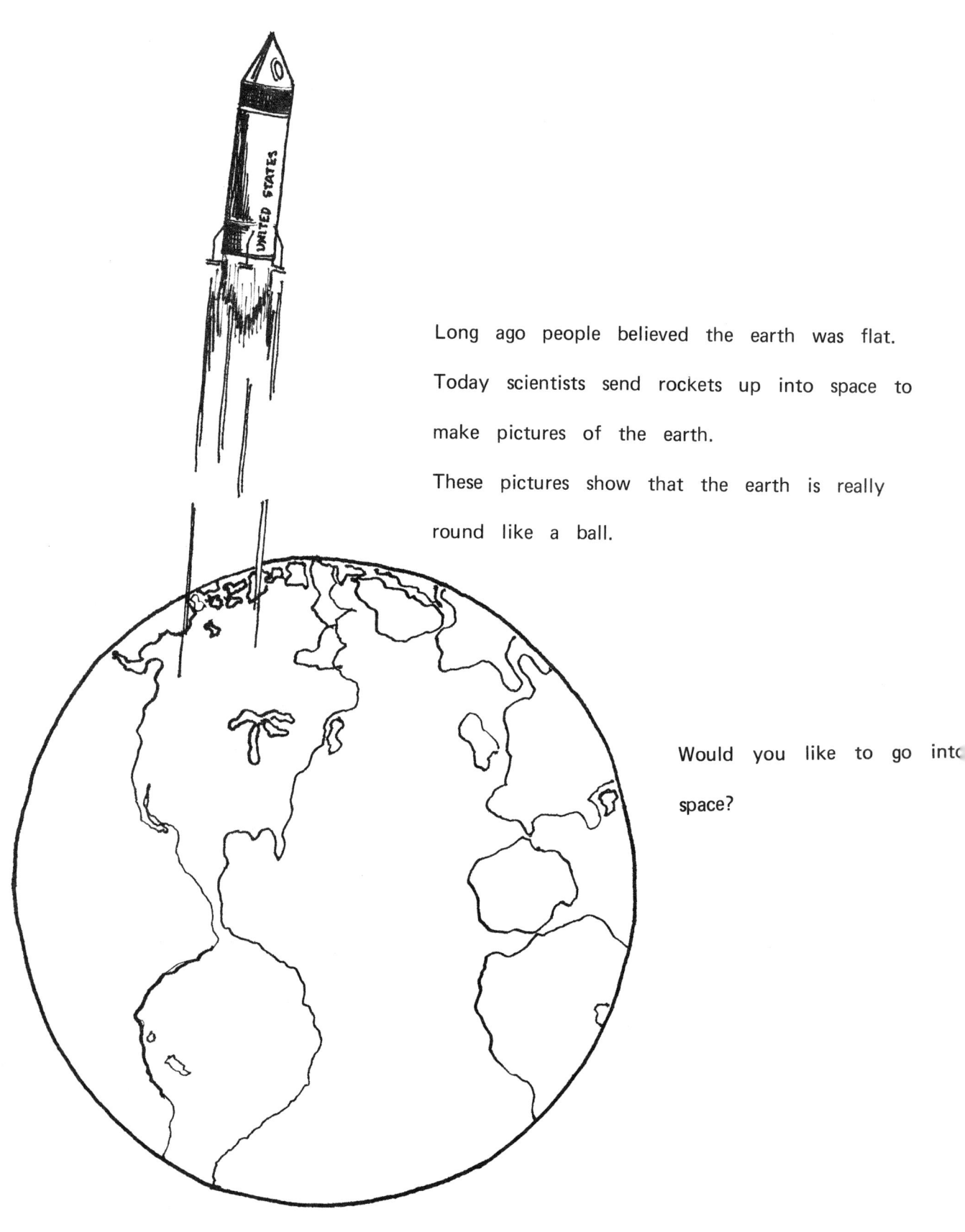

Long ago people believed the earth was flat.
Today scientists send rockets up into space to make pictures of the earth.
These pictures show that the earth is really round like a ball.

Would you like to go into space?

# What makes day and night?

The earth is turning slowly all the time.

As the part we live on turns toward the sun we have day.

As it turns away from the sun we have night.

Do you know how long it takes the earth to make one full turn?

Use your crayons to make this a daytime picture.

Use your crayons to make this a night time picture.

Here are two things you would use only at night.

Make a picture of one more thing that you would use when it is dark.

These two things would be used during the daylight hours.

Make a picture of one more thing you would use when the sun is shining.

Spring,
  Summer,
      Winter,
          Fall.

I like _____

          best of all!

Make a picture here to show the season you like best.

Use three colors in your picture

Apple blossoms are nice in spring.

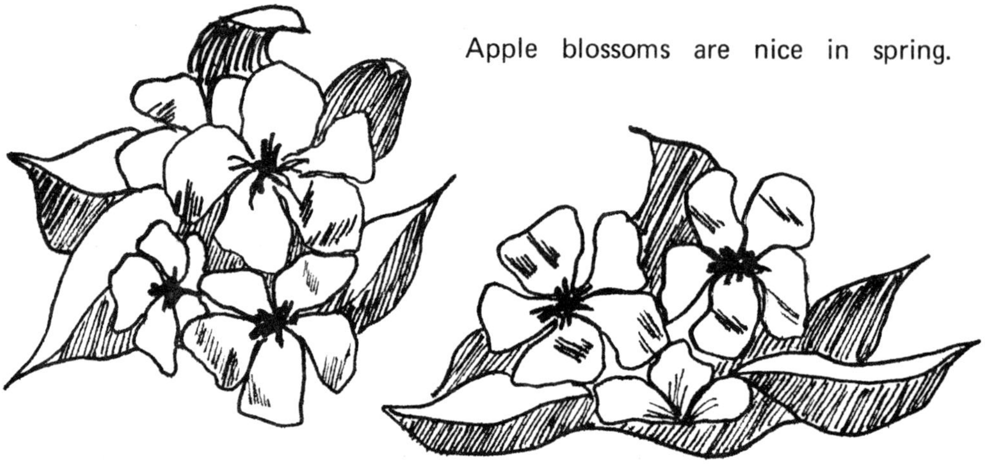

Picnics and parties are fun in summer.

Golden leaves are beautiful in fall.

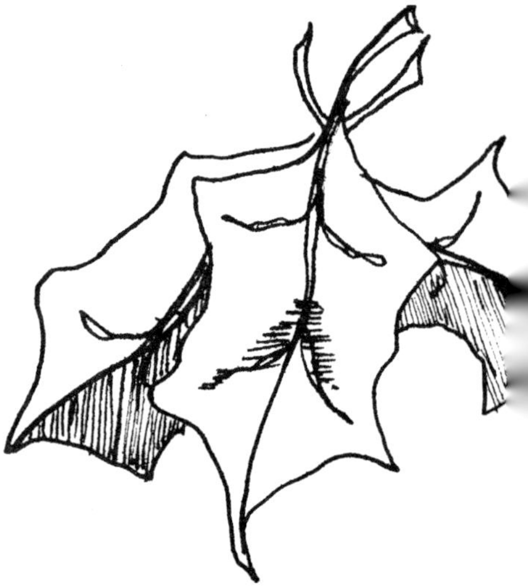

Snow and ice make a wonderland in winter.

Use your crayons to make these pictures show the seasons.

# What makes summer and winter?

The earth is *tilted* or slanted toward the sun so that the sun shines on it more directly at some times than at others.
When it is summer, the sun shines very directly on the earth.
When the sun's rays are not very direct, it is winter.
Spring and fall are in between.
Write the names of the four seasons.

spring

summer

fall

winter

# What do you know about the sun?

The sun is a star.

It looks <u>so-o-o</u> big because it is the star nearest the earth.

It is a daytime star.

Draw a picture of something you like to do when the sun is shining brightly.

The stars we see at night are shining in the daytime too.

We cannot see them until our part of the earth is turned away from the sun.

Do you know why?

# Have you ever wished upon a star?

Star light
 Star bright
  First star I see tonight
 I wish I may
  I wish I might
   Get this wish
    I wish tonight.

Have someone help you to write your wish here.

_____
- - - - - - - - - - - - - - - - - - - - - -
_____
_____
_____
- - - - - - - - - - - - - - - - - - - - - -
_____
_____
- - - - - - - - - - - - - - - - - - - - - -
_____

# What do you know about the moon?

Scientists have learned a lot about the moon by sending cameras inside space capsules to make pictures of it.

One of the things we have learned is that the moon has no wind or clouds.

We know too that the moon is hard enough for man to walk on.

How did we find this out for sure?

If you were a space man going to the moon, what three things would you want to take with you?

Draw a picture of them.

# Do you know how sand is made?

A sandy beach is made of many tiny rocks.

These tiny rocks that we call sand have been broken down

from big rocks by water and wind.

It takes many, many years for this to happen.

Make a picture of a sandy beach here.

Put yourself in the picture if you wish.

Long ago the earth was covered with big rocks.

Over the years the wind and water have broken lots of these big rocks into small rocks.

We can still find many large rocks if we look for them.

Make a picture of the largest rock that you can remember seeing.

Where did you see it?

# Here are three riddles for you.

Do you know what kind of rock we eat?

salt

Do you know what kind of rock we wear?

diamond

Do you know what kind of rock we burn?

coal

Did you know that you can learn about plants and animals that lived long ago by studying certain kinds of stones?

Some rocks have dents or prints made by animals or plants that lived and died long ago. These dents or prints are called "fossils."

Say this word so you will remember it.

FOR YOU TO DO . . .

Find some pictures of fossils in your encyclopedia or have someone draw some for you.

Take a ball of clay or play dough.

Make some make believe fossils to help you remember the word and its meaning.

Write the word.

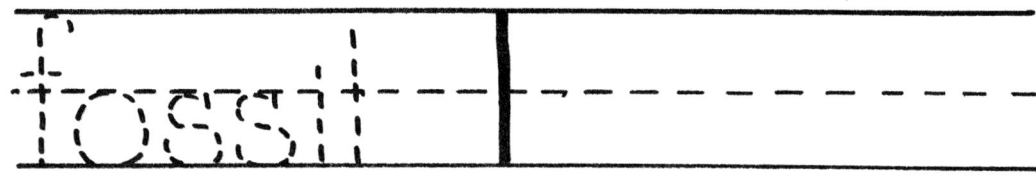

# Be A Rockhound

A rockhound is someone who collects and studies rocks.

Get a glass jar or a nice sturdy box for your rocks.

Ask someone to walk with you along a gravel road or beside a stream.

Look at all the rocks.

Take the most interesting ones for your collection.

Wash and polish them so you can see them more plainly before you add them to your collection.

Make a picture of your most interesting rock.

# Water and Air

# Where does the water go?

These pictures tell a story about water.

They tell what happens to water when it stands open to the air for a long time.

Picture 1 shows a glass of water almost full.

Picture 2 shows the glass the next day. Some of the water has disappeared.

Picture 3 shows that even more water has disappeared on the third day.

Where did the water go? Can you guess?

The water has gone up into the air!
Whenever water changes into vapor and goes up
into the air, we say it evaporates.

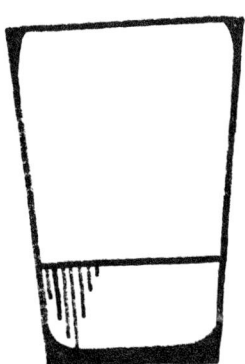

WOW! What a big word!

Can you say it?

Write it:

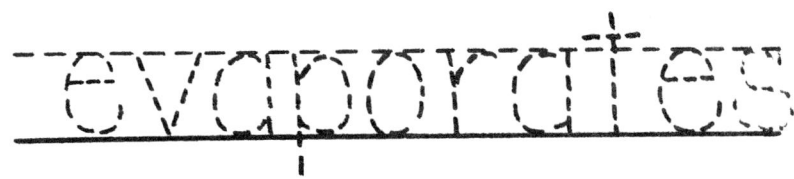

FOR YOU TO DO . . .

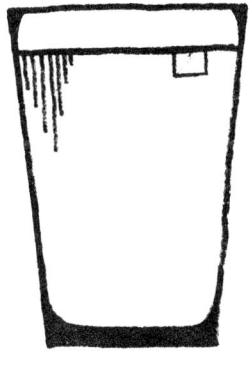

Fill a glass almost full of water.
Mark where the water comes with a rubberband or a piece of tape.

Let the glass sit for several days.

Each day, look at the water to see what has happened.

Does a little water disappear each day?

What word can you use to tell what the water does?

The water _____

# Can we cover the ice?

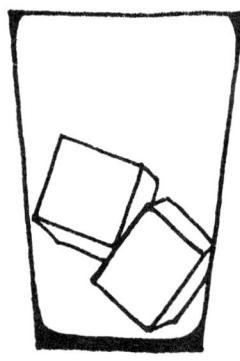

Here are some ice cubes in a glass.

Let's cover them with water.

Whoops! They won't stay under the water.

They keep coming up.

They float on top of the water.

Do you know why?

Ice cubes float on water because they are lighter than water.

# Which of these things will float on water?

Perhaps you can try some of them in a bowl of water.

Make a circle around the ones that float.

Why wouldn't the others float?

# Is dissolving the same as disappearing?

Here is a small glass of water.

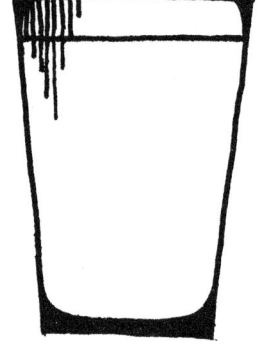

Let's put a spoonful of sugar in it.

Now we'll stir it.

Take the spoon out and look at the water.

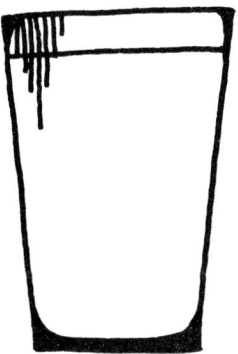

Can you see the sugar?

What happened to it?

It seems to have disappeared.

But if we taste the water, it will taste sweet, so we know the sugar is still there.

We say it has <u>dissolved</u>.

Can you say the word dissolve?

Write it.

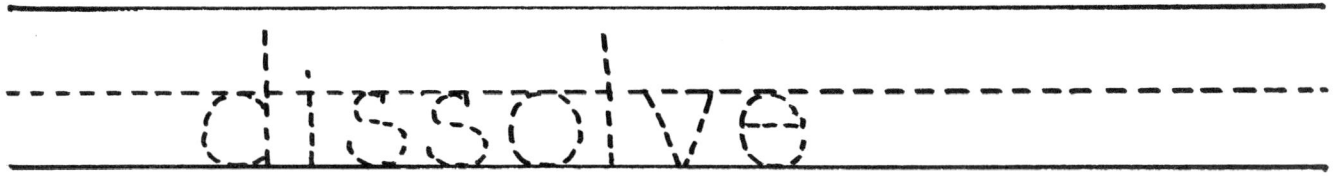

FOR YOU TO DO . . .

Dissolve a spoonful of salt in a glass of water.

Let the glass sit until all the water evaporates.

What is left?

water          add salt          stir

Let it sit.          What happened?

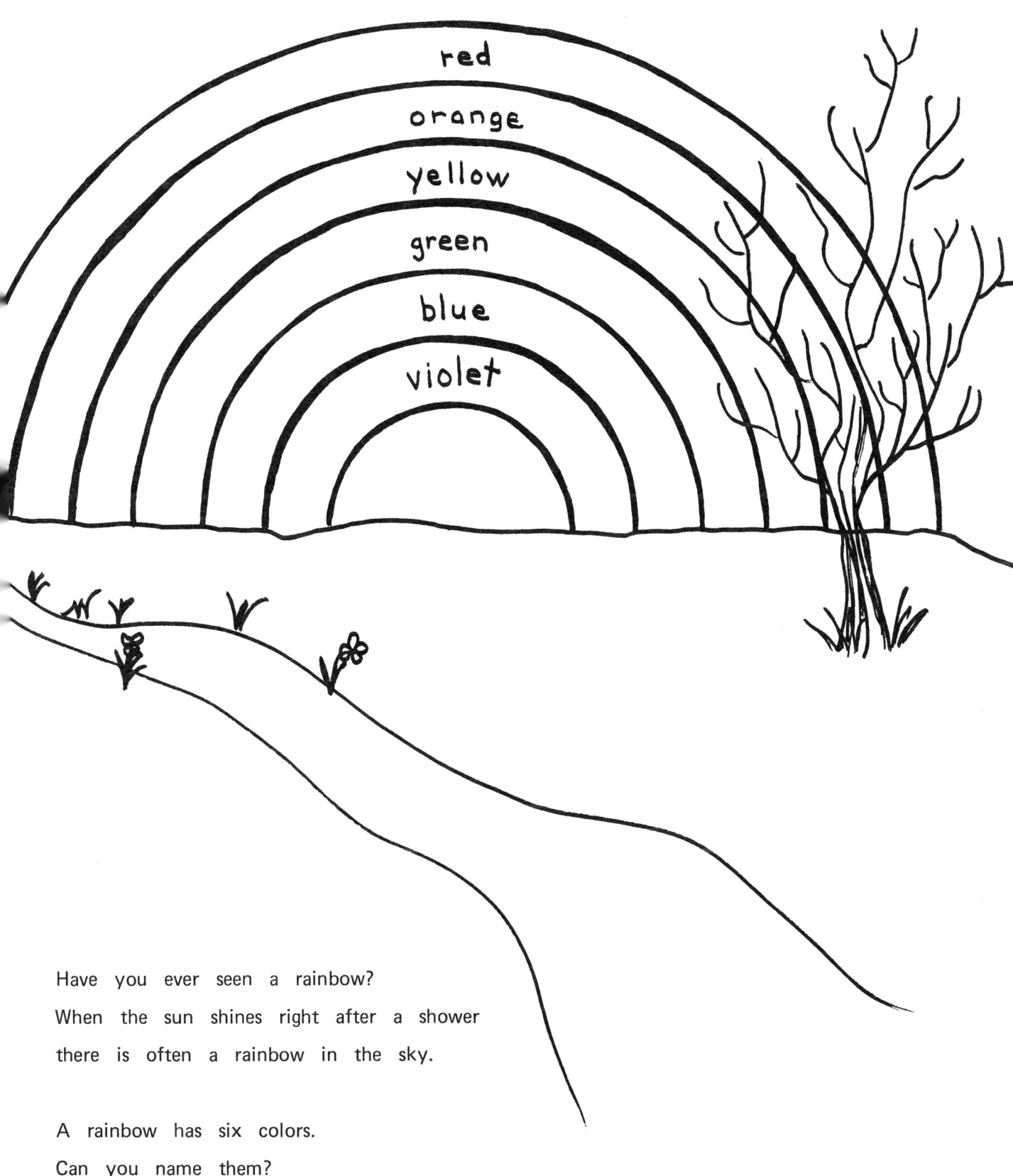

Have you ever seen a rainbow?
When the sun shines right after a shower there is often a rainbow in the sky.

A rainbow has six colors.
Can you name them?

Color the rainbow on this page.

# A RAINBOW PICTURE

by _____

Where is air?   Can you see it?

Can you feel it?   Can you hear it?

Air is everywhere.

It is all around you.

Put your arms out straight and whirl around.

Did you feel the air as it rushed by you?

Here are some experiments for you to do.

Do them - then see if you can answer the questions at the top of the page.

FOR YOU TO DO . . .

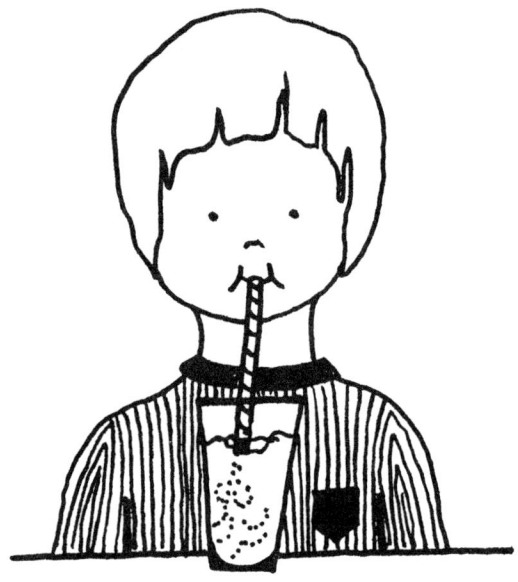

1. Find a drinking straw.

   Blow through the straw onto your hand.

   Blow through the straw into a glass of water.

   Can you feel air?

   Can you see air?

2. Blow up a balloon.

   Let it drop.

   What happened?

   What pushed the balloon around and around?

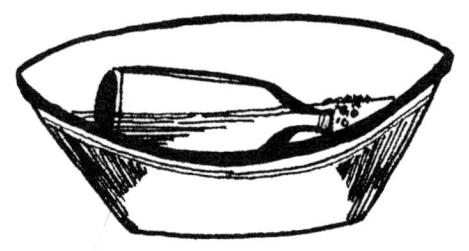

3. Put an empty Coke bottle into a big bowl of water.

   Watch the bubbles come up.

   What made the bubbles?

4. Place a tin can on a table.

   Try to blow it over.

   Can you do it?

   Now . . . fasten a balloon to the end of a drinking straw.

   Lay the balloon on the table and place the can on top of it.

   Blow through the straw into the balloon.

   What happened to the can?

Now can you answer these questions?

   Where is air?

   Can you see it?

   Can you feel it?

   Can you hear it?

# Machines, Magnets, and Electricity

# How do tools make tough work easy?

Can you name all of the things in the picture?

Can you tell how each is used?

All of these things are called tools.

Tools are simple machines that people use to make work easier.

This is a wheel.

It is one machine that helps people.

It would be very hard to drive a bicycle or a car without wheels.

Which picture below shows how wheels help people move?

Color it.

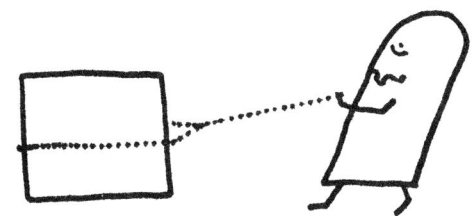

FOR YOU TO DO . . .

Here are pictures of some machines.

Can you find the ones that use wheels to make them work?

Color them and cut them out.

Paste them on the next page and add some of your own.

# TOOLS THAT USE WHEELS TO DO WORK

# Can you find some pictures of wheels in magazines?

Cut them out and paste them here.

# Which is heavier – the cat or the mouse?

Of course, the cat is heavier because he is much bigger.

But how does the mouse hold such a heavy cat up on the see-saw?

Could you lift your mother up off the floor in your arms?

Do you think you could lift her on a see-saw?

A lever is a machine that helps to lift things.

It makes lifting easier.

Which tools below are levers?

Circle them.

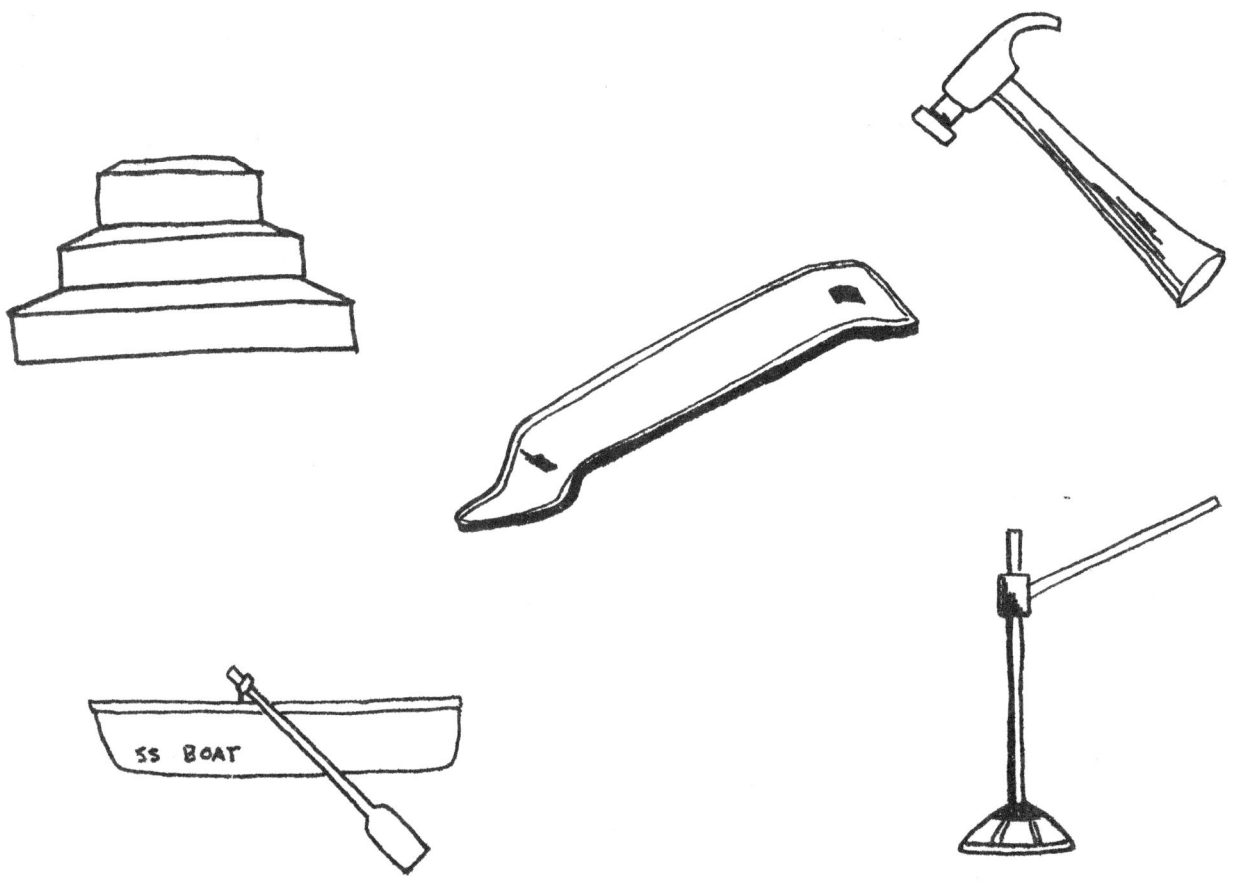

Do you think one small girl can balance two boys on a see-saw?

Yes, it is possible, because she has a machine to help her.

The see-saw is a lever.

The girl is standing on the long part of the see-saw.

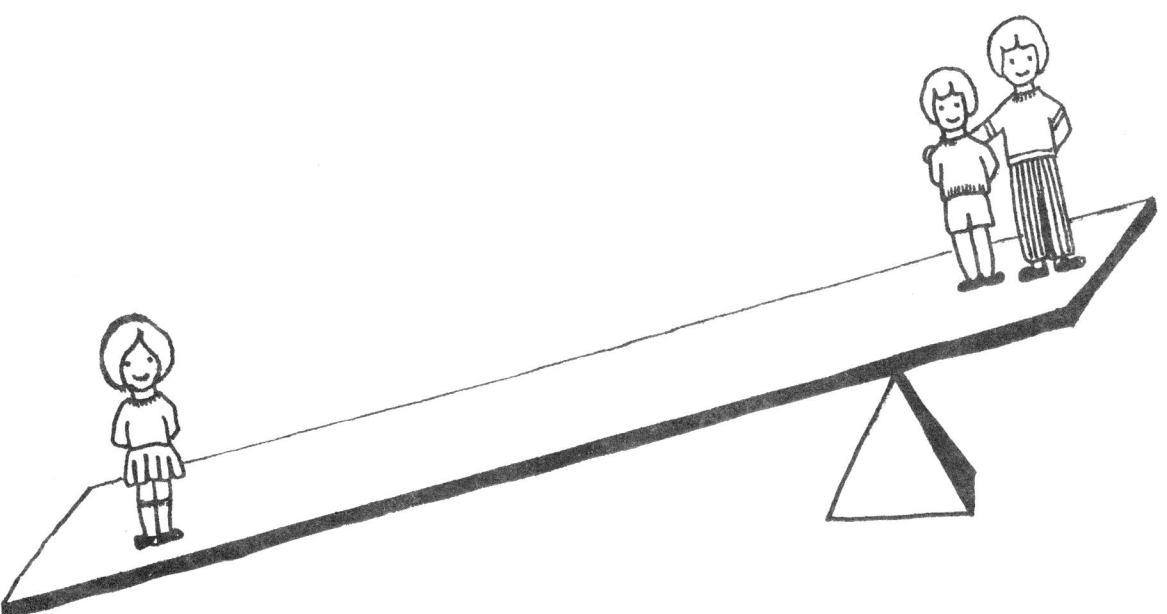

Because her side of the see-saw is longer and heavier, she can lift the boys.

You could lift your mother in the same way.

The long part of the see-saw on your side would help you.

FOR YOU TO DO . . .

Place a ruler on a small building block — like this:

Pretend it is a see-saw.

Place different objects on one end, and lift them up by pressing with your hand on the other end.

Can you lift things with the ruler?

What kind of machine is the ruler when you use it this way?

Write it.

lever

# Which truck will go faster?

This board is resting on a small stool.

If the toy truck is placed at the high end of the slope it will roll down.

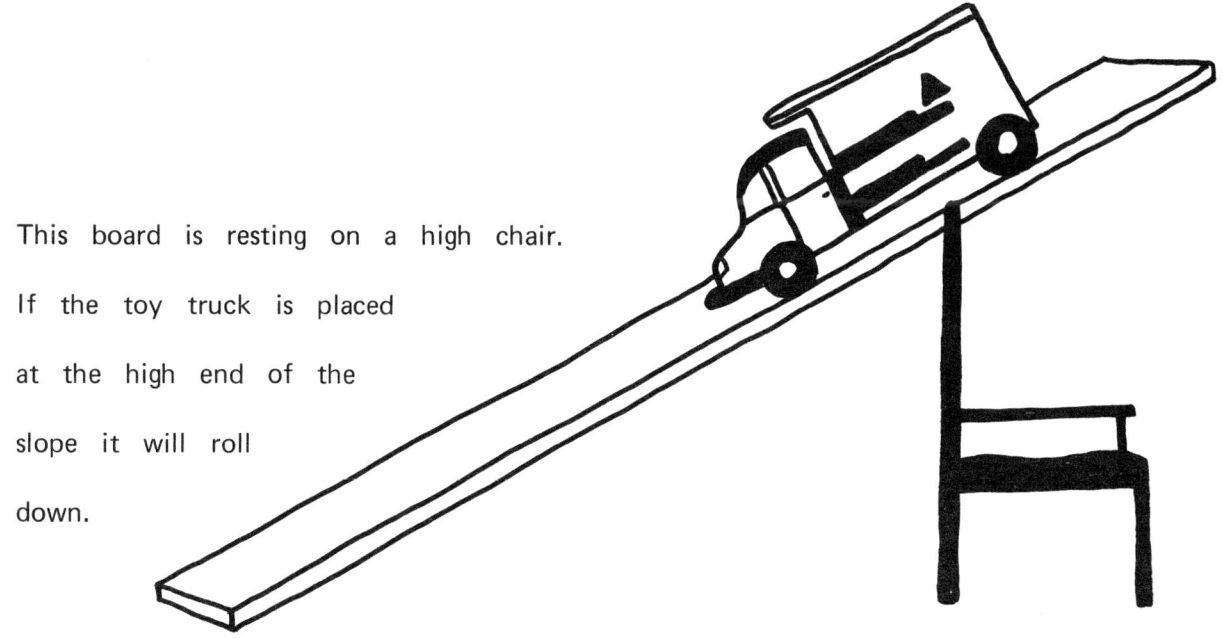

This board is resting on a high chair. If the toy truck is placed at the high end of the slope it will roll down.

Place an X on the truck which you think will roll faster.

If you said the white truck will roll faster, you are exactly right.

The board under the white truck is higher, and so the slope is steeper.

A board used as a simple machine is called an inclined plane.

An inclined plane makes going up or coming down much easier.

Write the words.

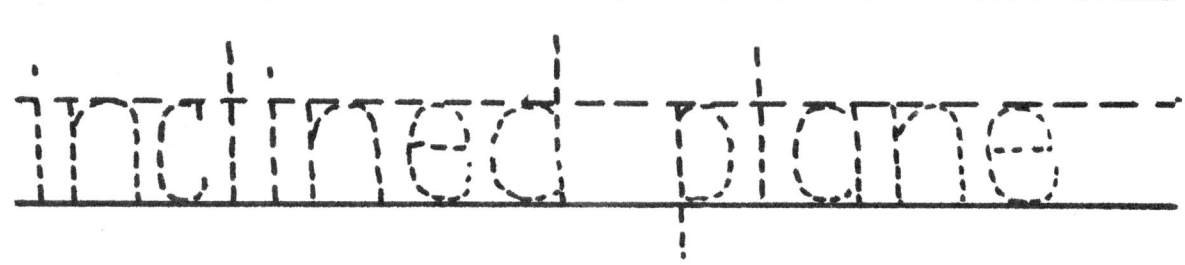

What a big step!
Only a giant could do it easily.

A boy or girl would not be able to get up on the landing without help. An inclined plane can help.

A stairway is an <u>inclined plane.</u>

Can you draw another inclined plane that will make the big step easier?

FOR YOU TO DO . . .

Ask a grown-up to find a heavy plank or board.

Place one end of the board on the edge of a chair or bench.

Let the other end rest against the wall.

Be sure someone holds the chair steady.

Step on the board near the wall and walk up to the chair.

Was it easy or hard?

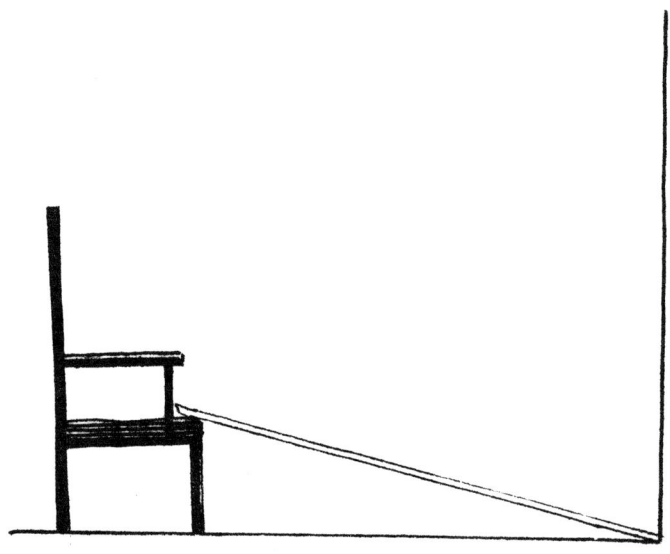

Now take the board away, and step up onto the chair by yourself.

Was it easier to use the board?

What machine made your work easier?

We have learned that simple machines make work easier.

Write the names of some machines below.

Make a picture to match each name.

There are many kinds of machines.

Color the machines on this page.

Put an X on the things that are not machines.

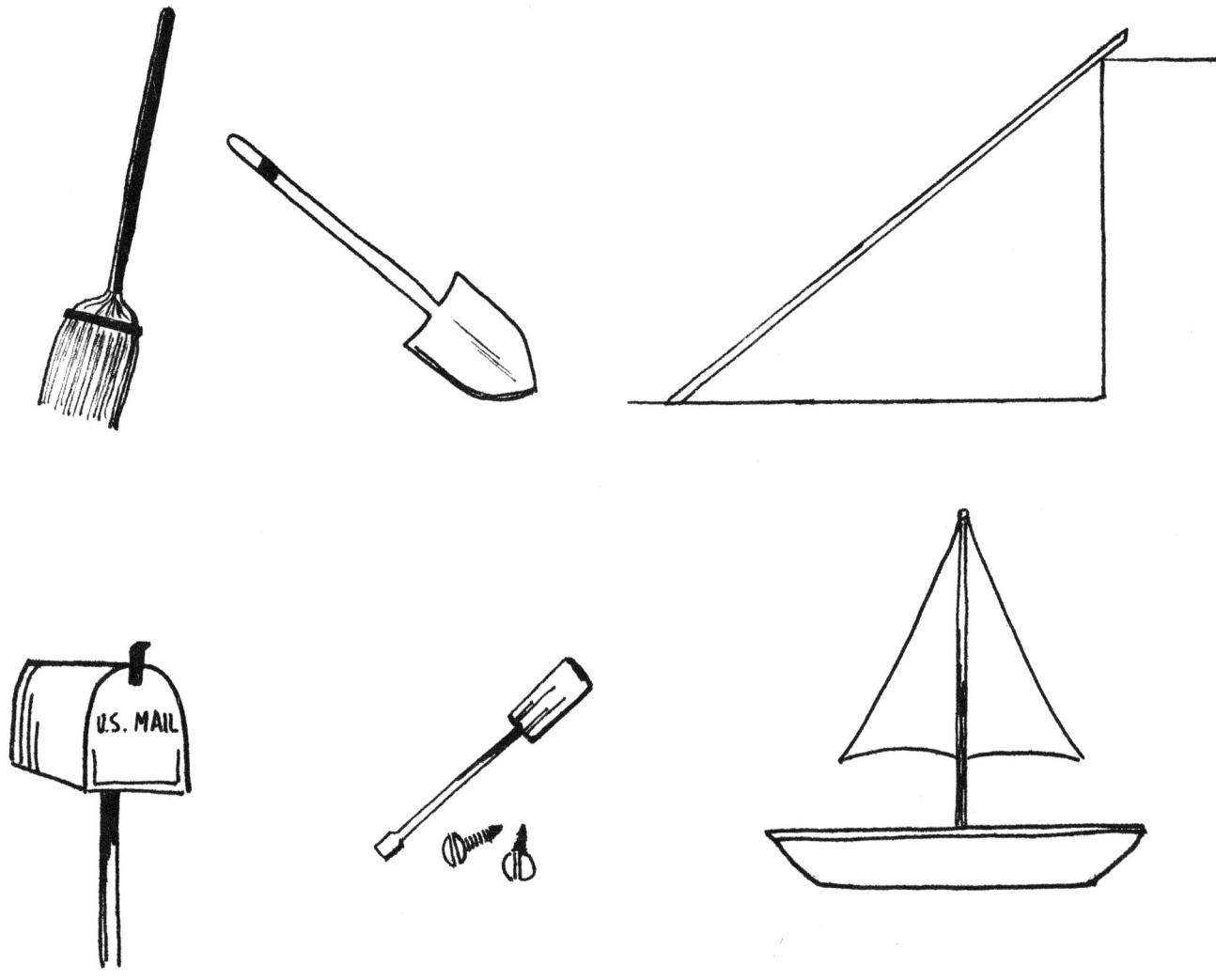

# Can you see electricity?

Electricity is all around us.

It is in the air and in the ground and in the things we touch.

But we can't see it.

We can only see what it does.

Look at the next page to see some things electricity can do.

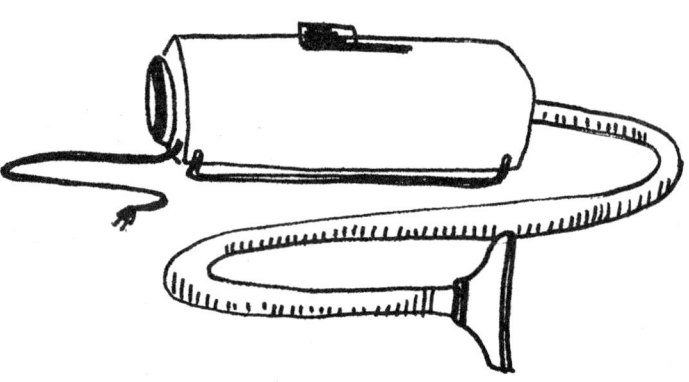
Electricity gives power for tools and machines

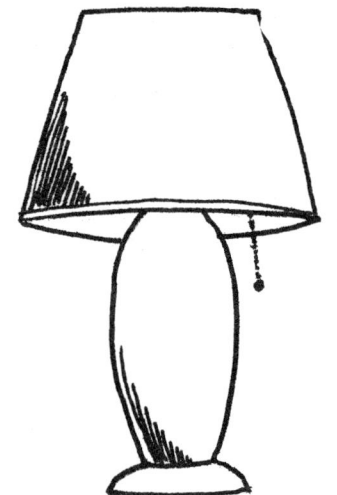
Electricity gives us light.

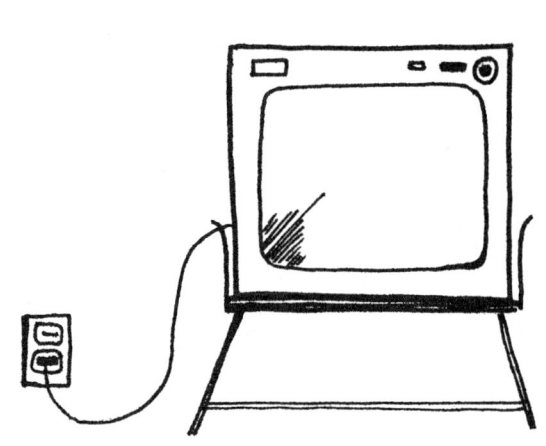

Electricity makes things move.

Electricity gives power for T.V., telephones, radios, telegraph and radar.

Electricity gives us heat.

# HOW ELECTRICITY HELPS ME

by _____

Electricity can make things work for children too.

Can you draw a picture of something electricity might do for you?

Can you tell which things get their power from electricity?
Circle them.

# Can you guess what these pictures are?

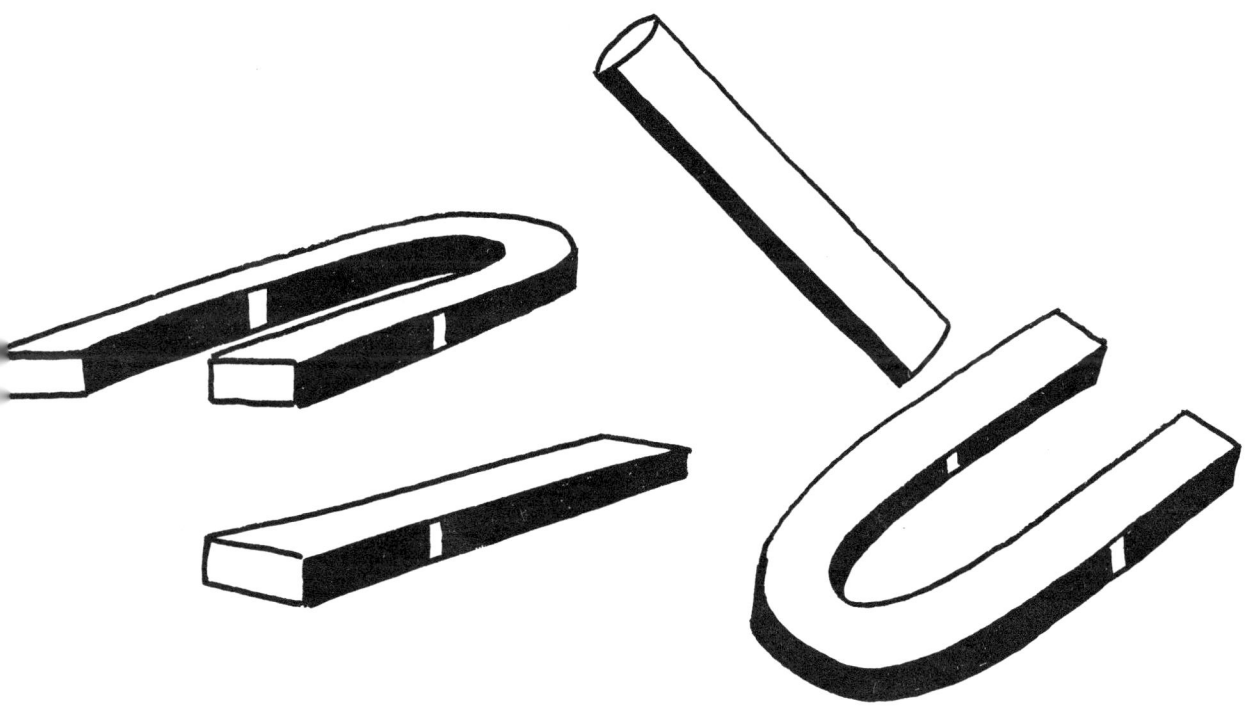

If you said they are pictures of <u>magnets,</u> you are exactly right.

A <u>magnet</u> is a stone or piece of metal which can pull certain things toward itself.

Count the pictures of magnets you see on this page.

Magnets can pull or <u>attract</u> or pick up only things that have the metals iron, nickel, or cobalt in them.

Which things below do you think a magnet could attract?

Make a red X on them.

# Is skin just a cover-up?

All animals have some kind of covering on their bodies.

People have skin.

Skin is wonderful, because it can do so many things.

Look at the skin on your body and think about these things.

1. It fits just perfectly!

(It isn't too big or too little, but just right!)

2. It is waterproof!

(If it weren't, you might soak up the water and sink in the bathtub!)

3. It keeps you at just the right temperature!

(It keeps your body warm when the weather is cold and lets you sweat to cool off when the weather is hot!)

4. It protects you!

(. . .from germs, from hot sun, from all kinds of harmful things.

It is thin where it needs to be, like on your eyelids - and thick in other places where you need more protection, like on your hands and the soles of your feet . . .)

How is your skin like some of the things in this picture?

How is it different?

# Where do freckles come from?

Almost everyone has something under his skin called pigment.

Pigment is color.

When you are out in the sun, the sun makes some of the pigment come out on your skin and you get a suntan.

Freckles are tiny spots on your skin where there is some extra pigment.

What are these things good for?

If you said, "Nothing", you are exactly right.

People do all kinds of funny things to make freckles or to get rid of freckles,

but freckles do just what they want to.

No one can make them come or go.

Here are some faces for you
to make freckles on.

Write freckles.

_____

- - - - - - - - - - - - - - - - - - - - - - - -

_____

Write the word that tells what makes freckles.

pigment

# Do toads give you warts?

This is a toad.

If you touched a real live toad, do you think you might get warts?

Some people think so, but they are wrong.

A wart is a little bump on your skin.//
It doesn't hurt and it often goes away by itself.//
Scientists think that warts may be caused by a virus on your skin.//
A virus is a germ too tiny to see.

# Ooooh! A skeleton!

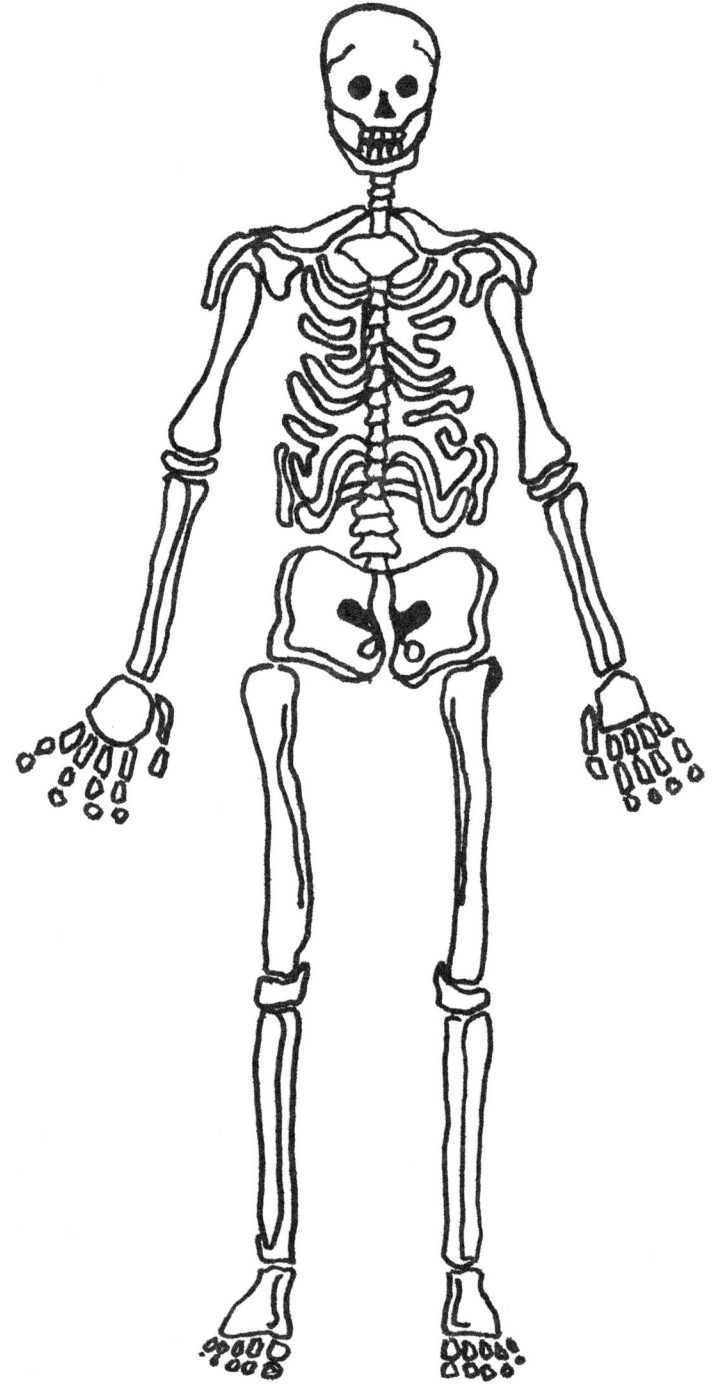

Does it look scary?  Of course not!  Skeletons are only scary on Halloween.

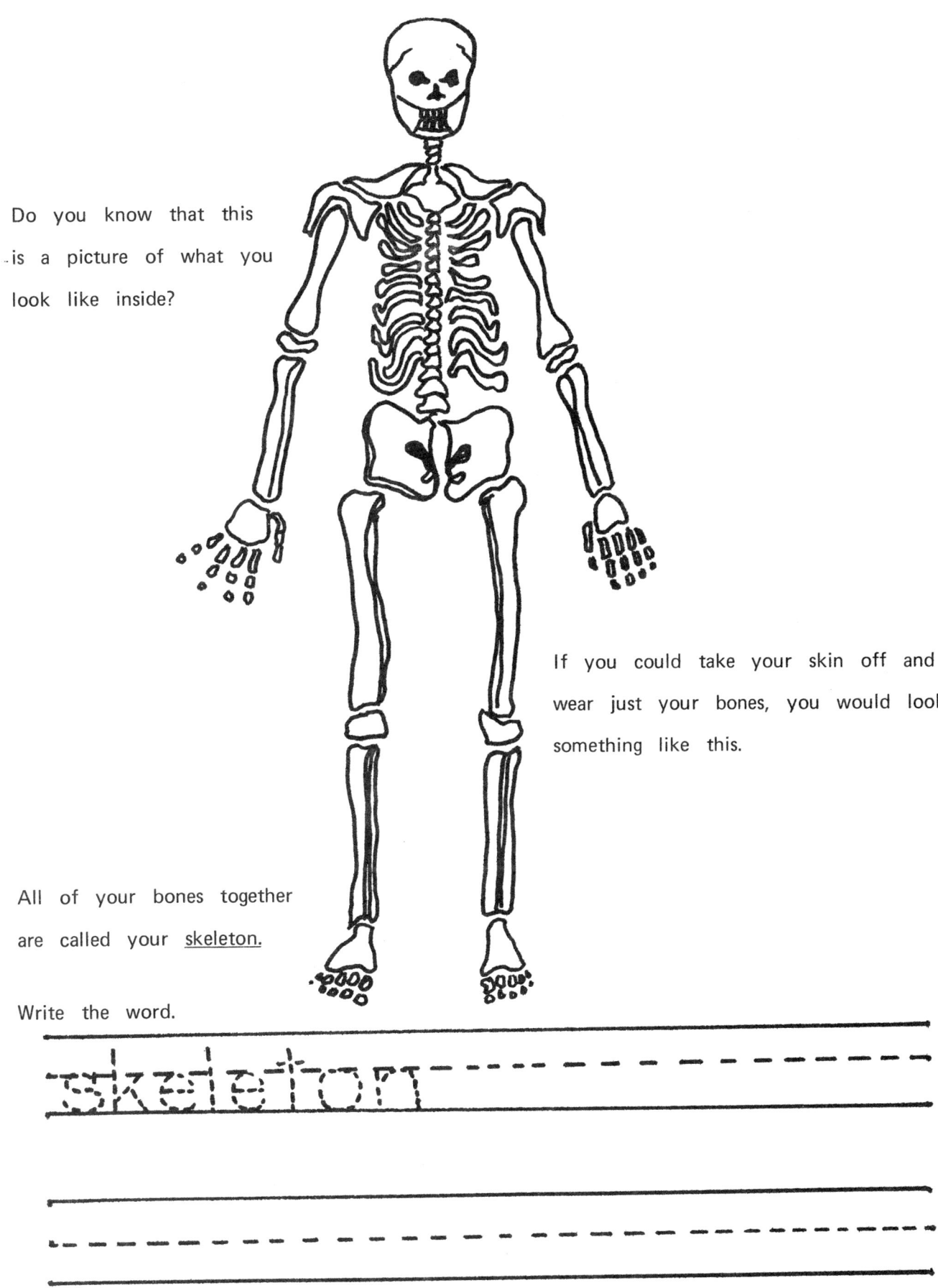

Do you know that this is a picture of what you look like inside?

If you could take your skin off and wear just your bones, you would look something like this.

All of your bones together are called your <u>skeleton.</u>

Write the word.

skeleton

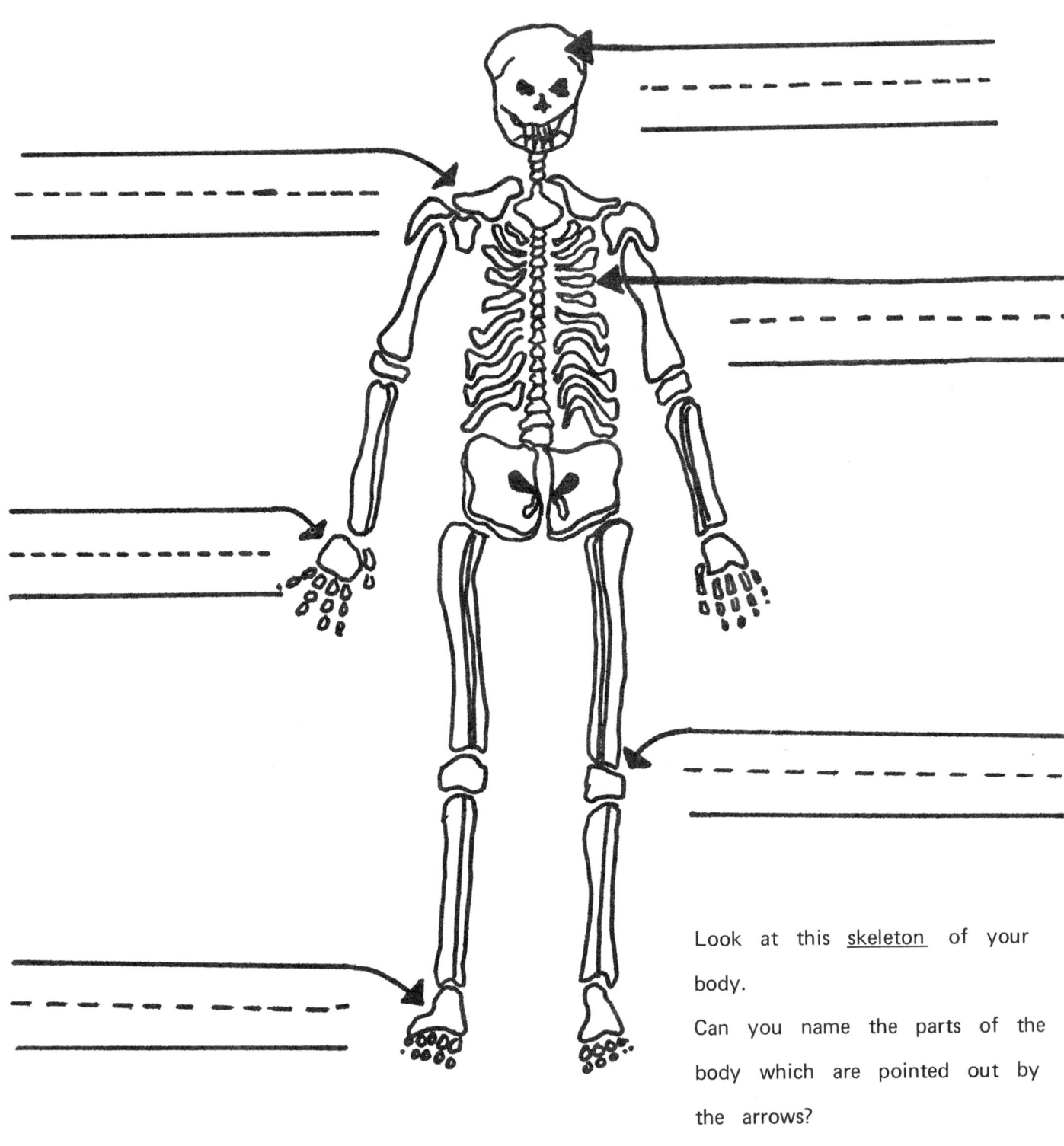

Look at this <u>skeleton</u> of your body.

Can you name the parts of the body which are pointed out by the arrows?

Use these words to help you label the parts of the body.

**head**  **shoulder**

**foot**  **hand**

**knee**  **chest**

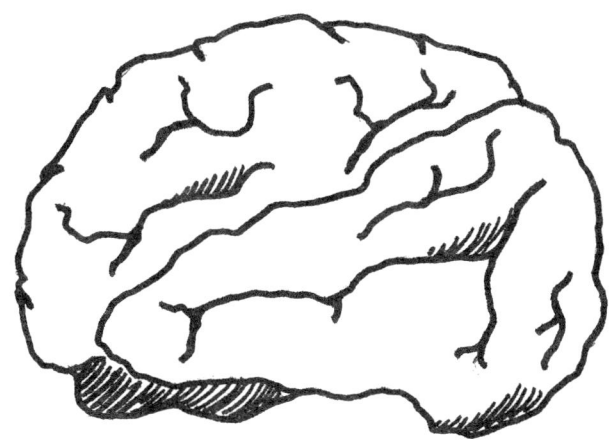

What does this picture look like?

You might think it is a sponge or a head of cabbage.

Would you believe . . . it's really a picture of a human <u>brain</u>?

Your brain is the very most important part of your body because it tells all the other parts when and how to work.

Your finger or eye or hand or leg — no part of you can move until your brain tells it to move.

Write the name of the most important part of your body.

_____
- - - - - - - - - - - - - - - - - - - - - - - - -
_____

 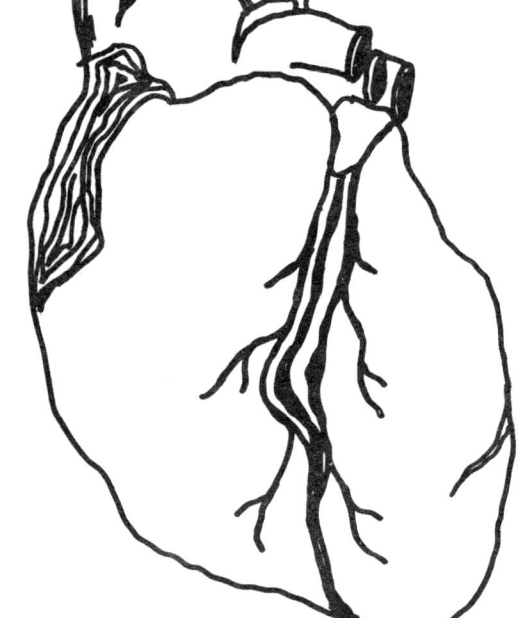

We call this a heart     BUT     if we could see inside a human heart, it would look much more like this.

The heart is a <u>muscle.</u>
It acts like a pump to move the blood through the body.  Each time it pumps, we can hear a <u>heartbeat.</u>
Your heart beats more than 100,000 times each day.
Have you ever listened to your heart beat?

FOR YOU TO DO . . .

Ask someone to show you how to feel your pulse.

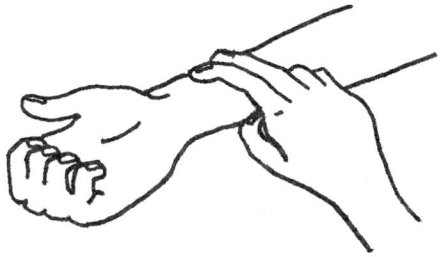

If you place your hand on your pulse, you can feel the pumping your heart does to move your blood.
Count your pulse while you are sitting still.
Then jump up and down several times.
Now count your pulse again.
What happens?

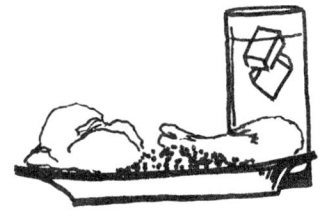  

FOOD                    REST                    EXERCISE

These pictures show some things human beings need to stay well and happy. Can you tell what they are?

A human being needs to take good care of his body so that all the parts of it will work right.

Here are some pictures of things he can do.

Can you think of some other ways you can care for your body?

Draw a picture of one way on the next page.

# TAKING GOOD CARE OF MY BODY

by _____

# Why do bodies do such strange things?

They sneeze!

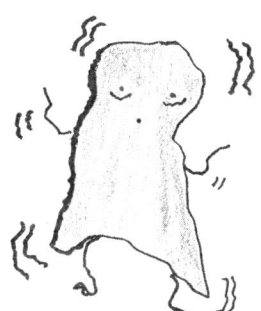

They shiver!

They sweat!

They yawn!

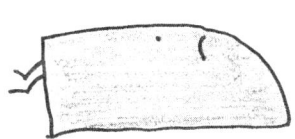

They sleep!

They hiccup!

Your body sleeps because it needs rest.

When it gets too tired to act the way it should, it needs to slow down.

When you sleep, all the working parts of your body can work more slowly, so they can rest.

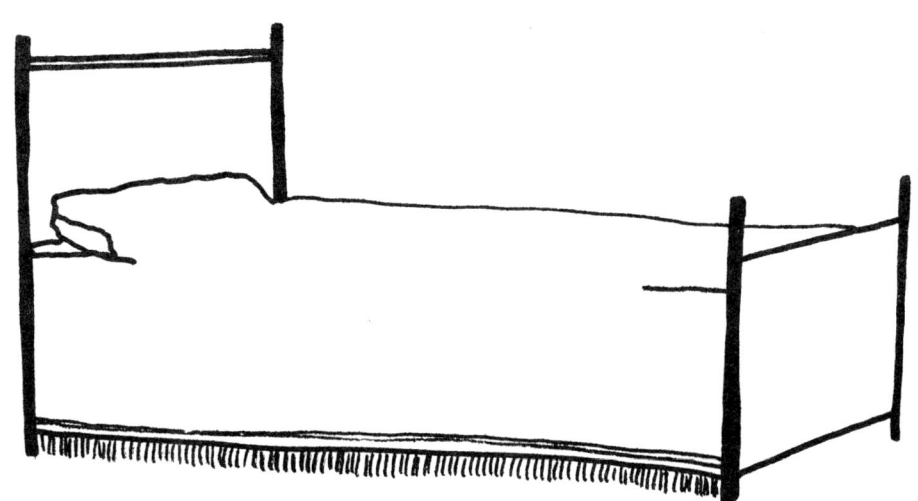

Put yourself to bed . . . draw a picture of you in the bed.

Do you know what else you do when you're tired?

You yawn.

Sometimes you yawn when you're not very tired.

Your body just needs some fresh air.

When you yawn you get a big gulp of fresh air, and breathe out a big gulp of old air . . .

Draw a picture of you yawning.

ACHOO!!

What happens to make you sneeze?

Sneezes sneak up on you, and you can't do much about them but let them come out.

You sneeze when something gets inside your nose that doesn't belong there. It may be a piece of dust or dirt. It tickles the nerves in your nose and makes you sneeze to blow the dirt out.

Here is a poem about sneezes. Say it . . . Teach it to a friend. It's fun to do!

### Sneezes

Sneezle - bambeezle - bumbozzle - KERCHOO!
You can't choose to sneeze
It just happens to you!

You can hop, you can flop,
But there's no way to stop
A sneezle - bambeezle - bumbozzle - KERCHOO!

When it's hot, my body <u>sweats</u> through tiny openings in my skin called <u>pores.</u>

When it's cold those pores close up.
And when I get very cold, the muscles in my body wiggle all by themselves to warm me up.
That is what <u>shivering</u> is.

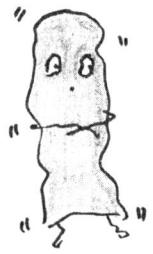

Is it hard to hiccup?

Whoops!   No!

It's hard to stop!

Do you know what makes you hiccup?

When you breathe, a muscle called your <u>diaphragm</u> pulls and pushes the air in and out of your lungs.

Sometimes that old diaphragm jerks and when it does it pushes a puff of air out past your voice box and makes you hic.

No one knows a sure way to stop hiccups.

People try lots of things.

Sometimes taking a long drink of water helps.

Have you ever tried to stop hiccups?

How did you do it?

# How does a body fix itself?

The body of an animal is one of only a very few things that can fix itself when it gets hurt.

When you cut your finger, you clean it, put a band-aid on it, and -

ABRA CA DABRA ! ! -

like magic — several days later you take the band-aid off and the cut has healed. Just a little more time, and you may not be able to see where the cut was at all.

AMAZING!   (That means it surprises you.)

But how does a body do it?

As soon as you cut your finger, it begins to bleed.

Tiny cells and fibers in the blood begin to stick together to make the blood thick.

Finally the blood gets hard and makes a cover or scab on the cut.

The scab protects the cut while other cells work underneath and inside your skin to heal the cut.

When the cut is healed, the scab falls off!

FOR YOU TO DO . . .

Which pictures make sense?

Talk about each picture with a friend.

Some of them are silly!

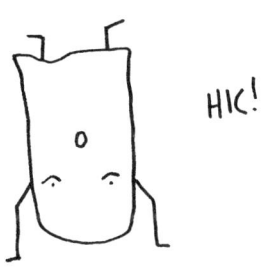

Which one is most sensible?

  . . . (later)    . . . (still later)

# Test yourself...... What have you learned?

Make an X on the picture that shows a <u>tool.</u>

Make a circle around the picture that shows a <u>mammal.</u>

Make a box around the animal that hatched from an <u>egg.</u>

Color the picture that shows <u>winter</u>.

Make a line under the picture of the object that uses <u>electricity.</u>

Make a dotted circle around the picture that shows <u>pigment</u>.

Make a circle around the picture that shows an <u>insect</u>.

Make an X on the picture that shows <u>night and day</u>.

Color the picture that shows a <u>magnet</u>.

Make a line under the picture that shows a <u>lever</u>.

Make a dotted circle around the picture that shows <u>seeds.</u>

When you have finished, compare your test pages with the answer pages which follow. They should look alike. Are you a good scientist?

# Answer pages

# Check yours ! .

# THIS IS

You have been studying about the world and your place in the world. Draw a picture of your world.

# MY WORLD

Be sure to include plants and animals, sky and earth, and some of the other things you have learned about in this book.
But most important of all, don't forget to draw YOU!

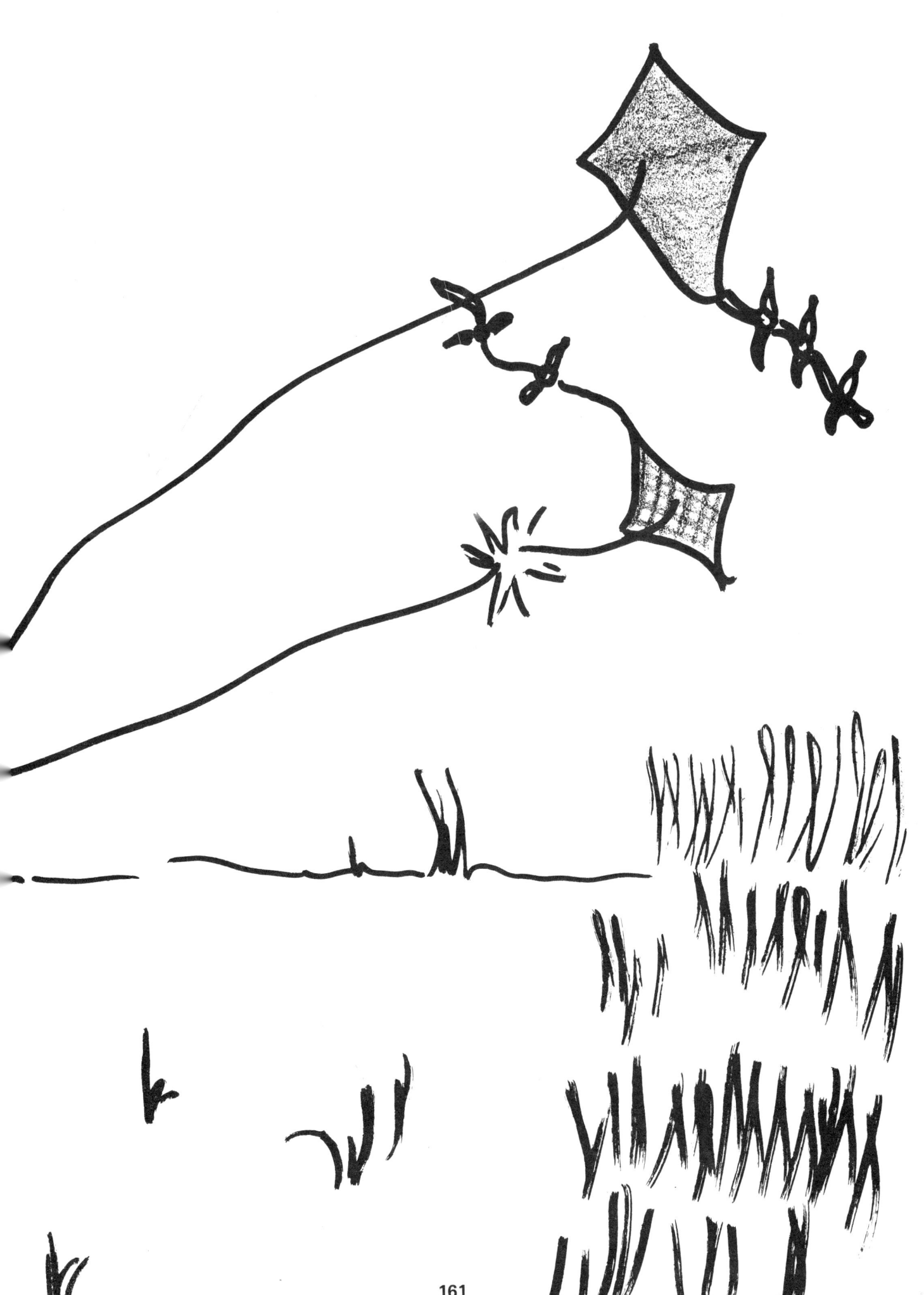